D0866523

The Secret Life of Space

The Secret Life of Space

HEATHER COUPER AND NIGEL HENBEST

First published in Great Britain
2015 by Aurum Press Ltd
74–77 White Lion Street
Islington
London N1 9PF
www.aurumpress.co.uk

A catalogue record for this book is available from the British Library.

ISBN 978 1 78131 393 0
eBook ISBN 978 1 78131 394 7

1 3 5 7 9 10 8 6 4 2
2015 2017 2019 2018 2016

Typeset in New Baskerville and Gill Sans by Saxon Graphics Ltd, Derby
Printed by CPI Group (UK) Ltd, Croydon, CR0 4YY

CONTENTS

Introduction

SHOULDERS OF GIANTS?

> 'If I have seen further, it is by standing on the shoulders of giants.'

So wrote Sir Isaac Newton to his colleague Robert Hooke on 5 February 1675.

And that's how we think of the discovery of the universe, from the days of the Flat Earth to the Hubble Space Telescope and the Big Bang. Great men have pioneered the way, with enormous insights and the overwhelming urge to change our view of the cosmos around us – think of Copernicus pursuing the idea that the Earth orbits the Sun; Galileo and his telescope; Isaac Newton and his theory of gravity; or Einstein with the mind-boggling concept of the Big Bang and black holes.

Except ... it wasn't really like that at all.

As we show in this book, most of the urban myths about astronomy are, in fact, just that – myths. Galileo didn't invent the telescope, and he wasn't the first to point the new instrument towards the heavens. Einstein didn't predict either black holes or the Big Bang: both were incarnated in the minds of almost-forgotten clergymen. Black holes were envisioned 200 years before Einstein was even born, and the great physicist didn't believe in either theory! Copernicus's theory that the Sun was central would never have seen the light of day if it hadn't been for a fanatical disciple of his, banned from his own country for being gay.

The poet Alexander Pope wrote:

> Nature and Nature's laws lay hid in night:
> God said, 'Let Newton be!' and all was light.

Yet Newton's laws would have remained hidden for all time if it hadn't been for an acolyte. His name was Edmond Halley: well known for 'his' comet, we owe Halley a much larger debt for pushing the older scientist to publish his *Principia*.

And then there are the unsung heroes. And the heroines: above, we intentionally used the phrase 'great men'. Yet many of the most important breakthroughs have been made by women, who – as a result of their gender – are even more overlooked than their male counterparts.

In the twentieth century, an Englishwoman from a small market town – Cecilia Payne-Gaposchkin – proved that most of the visible universe is made of hydrogen; but she was barred from taking a degree at Cambridge. American astronomer Vera Rubin mapped the invisible matter that controls galaxies, yet she too was ridiculed at international conferences.

Jocelyn Bell Burnell's untiring research in the 1960s unveiled the existence of pulsars; but she was overlooked for the Nobel Prize, which was awarded to her senior – male – colleagues. Though Caroline Herschel set a long-standing record for discovering new comets, she had to live in the perpetual shadow of her brother William, who found the planet Uranus in 1781.

William Herschel was a musician – and his discovery was a prime example of a major breakthrough made not by professional astronomers, but by amateurs who study the sky for the pure love of it. A thousand years ago, the Persian poet Omar Khayyam matched the Sun's movements to the calendar more accurately than anyone before or since. A nineteenth-century apothecary called Heinrich Schwabe discovered that spots on the Sun come and go every eleven years.

In our own times, a computer engineer has – from his backyard – found more exploding stars than anyone in history; a schoolteacher laid the basis for radio astronomy, and inspired the construction of the mighty Jodrell Bank radio telescope; while it was a sanitary engineer who has tracked down the strongest evidence yet for life on Mars.

There are also the mavericks – great professional astronomers 'forgotten' by the establishment because they didn't fit in. The Swiss astronomer Fritz Zwicky, an outsider in California who called his colleagues 'spherical bastards', first put astronomers on the trail of the 'dark matter' and 'dark energy' that scientists now think dictates the past and future of our universe.

Bluff Yorkshireman Fred Hoyle is mainly remembered today for proposing a rival to the Big Bang theory of the universe – the Steady State – which fell by the wayside. Yet Fred also led a team that made one of the most amazing breakthroughs of twentieth-century science, showing how stars are natural alchemists. In these cosmic cauldrons, the simplest element hydrogen is transmuted into carbon, oxygen and iron – and even the gold in your wedding ring.

In this book, we also explore new insights that overturn 'accepted wisdom'. For instance, Stonehenge was not built to worship the rising Sun on Midsummer's Day – as the 'druids' still do – but to observe the sunset on Midwinter's Day. A corroded lump of metal, found on the sea-floor in 1900, proves that the Greeks weren't just great thinkers: they actually built the world's first computer, to predict with amazing precision the movement of the Sun, Moon and planets – and even eclipses.

We explore the ongoing arguments that rage behind the astronomical scenes. Was it fair for astronomers to demote Pluto from being a planet? Is it bad – or good – when an asteroid hits our planet and wipes out most of the dominant life? And have astronomers been barking up the wrong celestial tree when – in their hunt for alien intelligence – they have focused on radio telescopes: maybe ET's phone is tuned to laser beams, or even to faster-than-light particles.

We also tell the inside story of the great cosmic adventures of our time, looking behind the scenes at the nail-biting saga of the *Rosetta* spacecraft and the little lander *Philae* as they probe at first hand Comet Churyumov-Gerasimenko, with their quest for the origin of life. And we follow the ongoing saga of the astronomers who have tracked down some 2,000 planets in the depths of space, in the search for an Earth-twin that could be our second home in the universe.

Our investigations into these largely untold sagas begins with the history of astronomy. Then we move into the realms of our neighbouring planets, the Sun and the stars. We dig deeply into the big mysteries of black holes and the universe, before finally exploring all aspects of life in the cosmos.

We'd like you to join us in celebrating the countless 'secret lives' that lie behind mankind's quest to unravel the mysteries of space, and hopefully think about them next time you sit in astonishment as a new image from the Hubble Space Telescope appears on your tablet or – even better – as you gaze up at the beautiful night sky.

Heather Couper and Nigel Henbest

PS: Here's another secret we can share. Don't take the quotation that leads this Introduction at face value. When he referred to 'the shoulders of giants', Newton was quite possibly directing an intentional insult at his great rival Hooke, who had developed a severe stoop in later life!

Chapter 1

SECRETS OF STONEHENGE

In the dark hours before dawn, the stars seem to crowd down on Stonehenge: a natural planetarium dome set on top of the immense stone arches that surround us. Stretched out – like a prehistoric sacrifice – on the central Altar Stone, we trace out the ancient constellations overhead. The stars of the mighty hero Hercules confront his primaeval foe, the celestial Dragon, wrapping its coils around the Pole Star.

It is the morning of 21 June 1984, and we're here to witness a re-enactment of the oldest and most sacred astronomical ceremony in the world: sunrise at Stonehenge, on Midsummer's Day. It's far from being a quiet mystical embrace of the cosmos, though. Campfires burn nearby, where devotees are celebrating all night long in makeshift tents; deep rhythms drum out, and our nostrils are assailed by the smells of seared food and exotic smoking materials.

Dawn whitens the sky. A melancholy chant begins to rise. A solemn assembly of white-robed figures processes to the heart of the hoary stones. The druids stare through the portal in front of us, towards the silhouette of the Heel Stone, an outlier to the main sacred circle. Their song grows to a frenzy at the moment of sunrise.

Sadly, this year, the magical moment is signalled only by the druid's digital watches: it's hazy, and we're not treated to the iconic sight of the dawning Sun sailing over the Heel Stone. But it doesn't stop the hippies from the makeshift camp invading the site. Midsummer revellers of all creeds swarm over the stones, climb to the top of the arches, and baptise their babies at the Heel Stone.

But are we actually here at the right season? According to a growing band of experts, Stonehenge was not built to honour midsummer. The new evidence reveals that the great stone edifice was a temple to the Sun at the opposite end of its yearly cycle, in December. It's one more controversy to add to the many that – throughout history – have dogged the enigma we call Stonehenge. But everyone agrees on one fact: the great monument is associated with the Sun.

To unravel the secret life of Stonehenge, we must gaze back far into the mists of antiquity. Its massive stones were raised long before we have written records. But there are tantalising hints in the travel guides written by ancient Greek and Roman scholars, suggesting a traditional worship of the Sun-god – whom they called Apollo – in an island far to the north of the known world. The historian Diodorus of Sicily, writing between 60 and 30 BC, says: 'There is also on the island both a magnificent sacred precinct of Apollo and a notable temple which is adorned with many votive offerings and is spherical in shape.'

Can this 'spherical' temple to the Sun-god, on a far northern island, be anything other than the striking circles of Stonehenge?

After the Roman chronicles, Stonehenge drew its enigmatic veils around itself again until the seventeenth century, when the kings of England took a personal hand in its story. King James I came across Stonehenge during his royal progress through Wiltshire in 1620, and commissioned the great architect Inigo Jones – who imported the Palladian building style from Italy – to examine the monument. In *Stone-heng: Restored* he convinced himself that the original design was classical, and Stonehenge must have been built by the Romans.

Thirty years later, Charles II visited Stonehenge in very different circumstances. Charles was trying to regain the crown after his father, Charles I, had been beheaded in the English Civil War, and he'd just lost the Battle of Worcester. In disguise, Charles fled to the south coast to take a ship to France. On the way, he spent an impatient day at Stonehenge. His companion recorded that Charles was 'reckoning and re-reckoning the stones, in order to beguile the time'. But it sparked a royal interest, and after Charles had returned to the throne in the 1660s, many curious scholars attempted to solve the puzzle of Stonehenge. The diarist Samuel Pepys noted that its stones were: 'as prodigious as any tales I ever heard of them, and worth going this journey to see. God knows what their use was!'

His colleague John Aubrey wasn't content just to wonder. Aubrey, a leading antiquarian, realised that most stone circles in the British Isles were far to the north and west – in regions untouched by Romans and other invaders, such as the Danes and Saxons.

They must have been constructed by an older race of indigenous Britons. So far, so good.

The Roman accounts named the priestly caste of ancient Britain as druids. Aubrey considered them savage people – 'salvage' in the language of his day – and described the druids as '2 or 3 degrees I suppose lesse salvage than the Americans'.

Despite his indictment of their character, Aubrey immediately concluded that the druids had constructed Stonehenge. The idea has taken a tenacious hold, right up to the present day. And it's entirely wrong. The Roman naturalist Pliny describes how druids 'choose forests of oak' for their sacred rites – no mention of stones – and we now know Stonehenge was two millennia old when the druids were (in Pliny's words) 'mumbling many oraisons and praying devoutly'.

While Aubrey introduced the druids into the Stonehenge saga, it was a fellow druidical convert who suggested the ancient priesthood erected the great stones specifically to observe the midsummer sunrise – another faulty idea that would cast its long and murky shadow over the entire future of Stonehenge.

William Stukeley was so infected with druidical fever that he gave himself the title 'Chyndonax the Druid'. But his interests were generally serious and wide-ranging, from coins to poetry, from music to science. He once dissected a dead elephant in London: the poor beast had been exhibited in public, but died because of 'the great quantities of ale the public continually gave it'!

Stukeley made exhaustive measurements at Stonehenge, and he noted something unusual about the direction of the 'principal diameter or groundline of Stonehenge, leading from the entrance, up the middle of the temple, to the high altar'. Looking back along this line, he deduced 'the intent of the founders of Stonehenge was to set the entrance full northeast, being the point where the Sun rises, or nearly, at the summer solstice'.

In 1740 Stukeley published his findings and interpretations in *Stonehenge: A Temple Restor'd to the British Druids*. At midsummer, he averred, the druids held their 'principal religious meetings or festivals, with sacrifices, publick games and the like'. From then onwards, Stonehenge and midsummer became inextricably entwined.

The enigmatic stone monument is also closely meshed with some of the most original thinkers Britain has produced. The tally was joined in Victorian times by the extraordinary Norman Lockyer.

Lockyer was a civil servant in the War Office. But his interests ranged widely: he was friends with the poet laureate, Alfred, Lord Tennyson, as well as the Pre-Raphaelite artists. After a neighbour loaned him a telescope, Lockyer began to study the Sun. In 1868, he found that giant prominences rearing above the Sun's surface are shining with light of a particular yellow hue that is not produced by any element on Earth. Lockyer concluded that the light came from an element that hadn't yet been found by chemists, and – after the Greek Sun-god Helios – he named it 'helium' (all-too familiar today as the gas that fills our party balloons).

Without climbing any official scientific career ladder, Lockyer became the world's first Professor of Astrophysics. Passionate about communicating science to the public, he founded the science magazine *Nature*, and pushed for the creation of the Science Museum in London. And thanks to his sporting interests in his few moments of leisure time, Lockyer co-authored *The Rules of Golf*.

While visiting temples in Greece, Lockyer's ever-fertile mind wondered why they were lined up in particular directions, and if that could be linked to the Sun. Travelling to Egypt, Lockyer surveyed the great temple at Karnak: 'beyond all question the most majestic ruin in the world'. He found that the main axis points towards the northwest, towards the Valley of the Kings – where the Sun sets at the summer solstice. On his return, Lockyer was inspired to visit England's own 'Sun-temple'. His calculations showed that someone standing at the centre of Stonehenge would see the midsummer Sun rise over the rugged outlying Heel Stone. 'Just as surely as the temple of Karnak once pointed to the Sun *setting* at the summer solstice,' Lockyer pontificated, 'the temple at Stonehenge pointed nearly to the Sun *rising* at the summer solstice' [his italics].

Midsummer's Day and the druids became inextricably linked in 1905, when hundreds of members of the Ancient Order of Druids (in reality, a secretive society founded in 1781) converged on Stonehenge for a mass initiation at the summer solstice – a tradition that's continued ever since.

But in 1965, our comfortable view of Stonehenge as a solar temple exploded. Instead, it seemed, the great stones were erected as a sophisticated astronomical observatory, for studying both the Sun and the Moon. It doubled as a powerful computer, for predicting eclipses – the most fearful of all sky-sights.

The claims were made in a book entitled *Stonehenge Decoded,* written by a British scientist working in

Boston, Massachusetts. Gerald Hawkins was using the new-fangled electronic computer for his astronomical research, and he decided to exploit its awesome number-crunching power to solve the riddle of Stonehenge once and for all.

For his readers' benefit, Hawkins described the appearance of 'The Machine'. The IBM 7090 consisted 'physically of about twenty units – tall cabinets filled with calculating and recording devices, many with two tape reels visible ... It requires about 45,000 volt-amperes of electric current, about 70 horse-power.' To show how times have changed, the smartphone in your pocket today is literally thousands of times more powerful than this behemoth!

Hawkins reasoned that the astronomers of old might have watched not only the Sun, but also the Moon, as it rose and set at important times of the year. Perhaps the megaliths of Stonehenge had been laid out to observe these celestial events? Using a detailed plan of the site, Hawkins programmed the IBM machine to work out the direction of lines that joined the myriad stones (and holes where stones had once stood). The computer calculated that an amazing number of the stone alignments did indeed indicate points on the horizon where the Sun and Moon rose and set at significant times.

'Those astonishing figures fairly haunted me,' Hawkins recalled. 'I wondered what the odds actually were against coincidence.' There's a mathematical formula to work out the probability of the stones lining up just by chance, but – as Hawkins admitted – 'the arithmetic would be horrible. I personally let the machine do the figuring.'

The IBM duly spat out the odds: less than one in ten thousand. Hawkins concluded that Stonehenge had been deliberately designed and built as a sophisticated observatory.

And that wasn't all. Surrounding the great stones is a mysterious ring of pits, identified and catalogued in the 1920s and named the 'Aubrey Holes' in honour of John Aubrey, inventor of the druid legend, who had noted five of these holes in his observations. Hawkins and his faithful computer showed how ancient people could have moved small stones around these holes, to work out when eclipses would happen: Stonehenge was also a computer.

Hawkins was ecstatic. Stonehenge had been designed 'for the joy of man, the thinking animal'. He saw a connection 'between the spirit which animated the Stonehenge builders and that which inspired the creators of the Parthenon, and the Gothic cathedrals, and the first space craft to go to Mars'.

The idea of a modern computer proving that Stonehenge was a Neolithic computer caused a sensation in the 1960s. Copies of *Stonehenge Decoded* flew off the shelves. Other scientists picked up the baton. A retired Scottish engineer, Alexander Thom, surveyed other stone circles throughout the land and found sophisticated observatories everywhere. The theory of the Stonehenge 'computer' was improved by Fred Hoyle, the greatest astronomical mind of the century.

But the archaeologists who'd devoted their lives to Stonehenge were far from convinced: one described Hawkins' book as 'tendentious, arrogant, slipshod, and unconvincing'. But how could they contradict the power of the computer? Among those best qualified to take up the challenge was a poacher-turned-gamekeeper. Clive Ruggles has been an astronomer, a mathematician and a computer expert. While still a student, he travelled to Scotland to check Thom's theory that lines of standing stones could have been prehistoric observatories – as we well remember, because we (and our large frame tent) formed part of the expedition!

Ruggles became a professional archaeologist, and then combined his diverse interests when he was appointed the world's first Professor of Archaeoastronomy at the University of Leicester. He's surveyed ancient astronomical sites around the world, from Hawaii to Africa.

'There were three fundamental flaws in Hawkins' calculations,' Ruggles discovered when he investigated the statistics of Stonehenge. Most importantly, Hawkins chose to measure 50 cases where stones lined up with one other, while in fact there are 111 different ways of sighting from one megalith to another in the massive stone labyrinth. So the chance of two stones lining up with an important point on the horizon are much higher than Hawkins' 'one in ten thousand'.

'When the flaws are corrected, the probability of chance occurrence increases to better than evens,' Ruggles concludes. 'In other words, the majority of the stones were placed for reasons quite unrelated to potential astronomical alignments.'

The other problem is that Stonehenge – unlike Hawkins' IBM computer – was not all built at once. The construction, and subsequent rebuildings, took place over a thousand years, so there can't have been a single creative genius behind Stonehenge. Some of the stones in Hawkins' Neolithic computer weren't standing at the same time as the others; and in other cases another stone would have blocked the view. Worst of all for Hawkins' theory, the 'eclipse-predicting' Aubrey Holes were dug – and then filled in again – before the main stones were erected. In other words, the supposed computer was destroyed before the Neolithic builders constructed the observatory that provided it with astronomical observations!

No one doubts the IBM machine did a good job. But its master instructed it to look at the data in a faulty way. As computer geeks put it: 'garbage in, garbage out.' Today, the theory of the Neolithic computer has been logged off. It was a grandiose idea, but failed to live up to scrutiny.

'Stonehenge was a huge temple,' Ruggles concludes, and a major ceremonial site, with some of its megaliths coming from as far away as Wales. 'By building this powerful monument that includes stones from exotic places,' he continues, 'it places Stonehenge very forcefully in the cosmos as the centre of things.'

While the early astronomers may have been interested in the Moon, Ruggles believes their main concern was undoubtedly the Sun. 'I have no doubt that the solstitial alignment was part of the significance of that temple,' he says. 'But there are several reasons to suppose that it was actually the opposite direction – towards midwinter sunset – rather than the view outwards towards midsummer sunrise, that was most significant.'

Once Hawkins had opened the Pandora's Box of alignments at Stonehenge, it's not surprising that archaeologists have looked again at the long tradition that associates it with Midsummer's Day. While you can indeed attempt – as we did in 1984 – to watch the sunrise over the Heel Stone in June, you can equally well stand beside the Heel Stone on Midwinter's Day and look the opposite way: through the centre of Stonehenge, to see the Sun setting.

So, why did midsummer become entrenched in the lore of Stonehenge? It all goes back to William Stukeley. He declared in 1740 that Stonehenge was designed with its 'entrance' in the direction of sunrise at the summer solstice. Once that hare had been set running, everyone else followed it blindly – right up to the druids of the present day.

And yet, if we go back to Stukeley's original account, he began by describing the main axis of Stonehenge as 'leading from the entrance, up the middle of the temple, to the high altar' – which is the direction of midwinter sunset!

Why did he change his mind? The true reason – like so much else concerning Stonehenge – is long lost to us. But there is one clue as to why Stukeley picked on the midsummer sunrise, with its compass bearing of northeast. As well as being a self-proclaimed druid, Stukeley was a Freemason. To this day, the east – the direction of the rising Sun – is all-important to the members of a lodge. Writing two hundred years after Stukeley, the Freemason Harry LeRoy Haywood stated in his book *Symbolical Masonry*: 'If there is one symbol that recurs again and again in our Blue Lodge Ritual, like a musical refrain, it is the East ... the Blazing Star shines there ... it is the bourne, the goal, the ultimate destination, towards which the whole Craft moves.'

Particularly important is sunrise on Midsummer's Day. It's celebrated by Freemasons on the Christian feast day of St John, which is 24 June (astronomically, the date has slipped to 21 June because the early calendar was a bit wonky, as we discuss in Chapter 3). The first Grand Lodge of this secretive organisation had met on St John's Day in 1717 – only a few years before Stukeley's investigations at Stonehenge.

No wonder Stukeley looked to the direction of midsummer sunrise, and ignored the bearing of midwinter sunset: to Freemasons, this was the direction of death. In Haywood's words: 'To the West all men come at last, men and Masons, to the beautiful, tender West, and lay them down in the sleep that knows no waking.'

From his time spent worshipping in Christian churches, the scientist Norman Lockyer was bedazzled by the same direction: 'in England the eastern windows of churches face generally to the place of sunrising on the festival of the patron saint; this is why, for instance, the churches of St John the Baptist face very nearly north-east.'

But there's no reason to think that our Neolithic ancestors felt the same way. There are two great tombs in the British Isles which are definitely aligned with the Sun at the solstice, and where there's no doubt about which season was meant.

Acting on a hunch, Irish archaeologist Michael O'Kelly huddled deep inside the great burial mound of Newgrange, in the Boyne valley north of Dublin, early one morning in 1967. Then something miraculous happened. A shaft of brilliant light from the rising Sun suddenly shone straight up the long entrance passageway and into the core of the dark tomb. 'I was literally astounded,' recalled O'Kelly. 'There was so much light reflected from the floor that I could see the roof twenty feet above me.'

It was Midwinter's Day – the only time of the year when the Sun penetrates into Newgrange. Similarly, in Orkney, the Sun shines up the passageway of the Maes Howe burial mound – this time at sunset – only at the midwinter solstice.

Though we can't look into the mind of Neolithic people, there's a good parallel to be found at Chaco Canyon, New Mexico. Ruins of huge buildings litter the canyon floor – accommodation for thousands more people than could actually live in this arid desert. Many of these structures line up with the Sun's position on the horizon at either midsummer or midwinter: but which?

The stones here – as at Stonehenge – are mute. At Chaco Canyon, however, we find direct descendants of the builders still living in villages atop the nearby mesas. Each Pueblo village has a Sun-chief, whose job is to follow the Sun around the horizon during the year. And the focus of the solar year is midwinter.

'Many of the Pueblan ceremonial times revolve around the winter solstice,' explains local guide G.B. Cornucopia, 'to do the proper ceremonies so that the Sun will come back into the sky and start another yearly cycle.'

It makes a lot of sense. All autumn long, the Sun has been sinking, and the days dwindling. It evokes a primaeval fear that the life-giving orb might disappear entirely. Cornucopia explains that the now-ruinous 'Great Houses' were built to house hordes of people from all over the American Southwest, who came to join in the midwinter rites.

New discoveries show that Stonehenge played a very similar role in England during Neolithic times.

Archaeology has given a voice to the long-deceased builders of Stonehenge, declaring that they travelled here, too, to celebrate in the depths of winter.

Archaeologist Mike Parker Pearson has investigated not only Stonehenge, but the nearby Neolithic sites and the landscape that surrounded them in prehistoric times. His team has discovered the remains of a huge settlement at Durrington Walls, in the valley below Stonehenge. 'It's the largest Neolithic village found in Britain,' says Parker Pearson, 'and it probably housed the people who built Stonehenge.'

And Durrington Walls was the site of great celebrations. Parker Pearson's team has uncovered the remains of orgiastic feasts: the bones of cows and pigs, and broken pots, some still containing traces of stew. The animals had been brought huge distances – some from Devon or Cornwall, and others perhaps even from Scotland.

Like Chaco Canyon, Stonehenge was a magnet for people from a whole nation. And the bones reveal that people gathered at the 'wild town next to Stonehenge where the builders partied' – in the words of a newspaper headline – at the same time of year. The growth of the pigs' teeth, and the amount they had worn, showed that the vast majority of these animals had been slaughtered for the table at the age of nine months. Given that pigs naturally give birth in the spring, 'people were here, feasting on pork,' concludes Parker Pearson, 'at midwinter – most likely around the midwinter solstice.'

In his book *Stonehenge*, Parker Pearson also explains why the country's greatest monument was constructed on this particular site. Beyond the Heel Stone, his team has found a set of strange features etched into the rocky surface that lies beneath a thin covering of soil. This set of long thin gullies was created naturally during the last ice age, thousands of years before Stonehenge was even conceived.

'We had stumbled upon the reason why Stonehenge is where it is,' Parker Pearson enthuses. From a natural mound, the 'periglacial stripes' line up precisely with the position where the Sun dropped below the horizon on the winter solstice.

'For Neolithic people this was where the passage of the Sun was marked on the ground,' he continues, 'where heaven and Earth came together.' It was so important that the Neolithic people imported giant stones to this site, some from a great distance. They erected the stones at the far end of this natural landing strip for the Sun.

'There, facing the midwinter sunset,' says Parker Pearson, 'they constructed a circular enclosure, whose roundness echoed that of the Sun and Moon.'

The idea that Neolithic people gathered here at the winter solstice, conducting rituals to assure the Sun's safe return, is the latest in a long line of theories – albeit one more soundly based in science – that have purported to explain what motivated our ancestors to create the most superb of ancient monuments. Over the centuries, the interpretation has reflected the prevalent zeitgeist: classical Roman temple, druid sacrificial site, cathedral to the Sun, sophisticated observatory, Neolithic computer – and now a ritual centre to ensure the return of sunny and prosperous days.

Stonehenge is an ink-blot test for the concerns and beliefs of the current age. As the leading British archaeologist Jacquetta Hawkes eloquently wrote: 'every age has the Stonehenge it deserves – or desires'.

For the moment, there's pretty overwhelming evidence that the great stone edifice was indeed a temple to the Sun, but not at the time of year when most people celebrate. Today, if we were to return to relive the experiences of our Neolithic forebears, it won't be a pleasant vigil, nested inside the stone circle on a warm summer's night. Instead, we'll be shivering on a winter's afternoon, watching over frosty fields as the dying Sun sets on Midwinter's Day. Who knows; the weather might even be better!

Chapter 2

THE GREEK COMPUTER

You can't help but be overwhelmed by the riches of the National Archaeological Museum in Athens. Spectacular statues, tombs and jewellery fill gallery after gallery in a dazzling exposition of one of the world's greatest civilisations.

So you could easily overlook a case that contains a corroded bronze wheel, no larger than a saucer, along with smaller lumps of broken metal. But the Antikythera Mechanism is the most precious item in the entire museum. More than 2,000 years old, it's the world's first working computer.

This device was built – like NASA's latest number-crunchers – to predict the motions of the heavenly bodies. The great science writer Arthur C. Clarke wrote: '… the Antikythera Mechanism represents a level that our technology did not match until the eighteenth century, and must therefore rank as one of the greatest basic mechanical inventions of all time.' Found in an ancient shipwreck off the island of Antikythera, it has single-handedly overturned the traditional idea that while the ancient Greeks were brilliant with their minds, they were totally impractical – the original 'absent-minded professors'.

Take the 'father of science', the early philosopher Thales. As he walked home one night – around 600 BC – pondering the mysteries of space, Thales fell into a ditch. An old lady helped him out, with the words, 'How can you hope to know anything about the stars above, when you don't even know the Earth beneath your feet!' His ruminations had one positive outcome, though. Thales realised that the Sun, Moon and planets were not gods and goddesses parading across the sky. Instead, they were inanimate objects. And the heavenly bodies had to obey natural laws.

In contrast, the Egyptians of the time were conducting rituals all night long, to persuade the Sun-god to get out of bed in the morning. And the astronomers in Babylon tracked the motions of the planetary sky-gods, to try to predict what these deities had in store for them.

Following Thales' lead was Pythagoras, who founded a secret school in the south of Italy. His followers had to give up their personal possessions and eat a purely vegetarian diet. Unusually for the time, Pythagoras treated his female pupils on the same terms as the men. The famous Pythagoras's Theorem was just one example of his obsession with numbers. Pythagoras looked for them everywhere – even in the musical note that rang out when a blacksmith hit his anvil. He discovered that the bigger the anvil, the lower the note it sounded. Inspired by the harmonious blacksmith, Pythagoras founded the theory of harmony that underlies the length of the strings in your piano today.

If devices on Earth could make sweet music, Pythagoras was convinced that the heavenly bodies must be even more melodious. The Moon, Sun and planets travel round the heavens at different rates and so, he theorised, they must 'sing' at different pitches. Pythagoras called the notes of this natural heavenly choir 'the Harmony of the Spheres'. Sadly, humans are too flawed to hear the sound. This idea resonated down the ages. Shakespeare refers to the Harmony of the Spheres in *The Merchant of Venice*:

> There's not the smallest orb which thou behold'st
> But in his motion like an angel sings,
> Still quiring to the young-eyed cherubins;
> Such harmony is in immortal souls;
> But whilst this muddy vesture of decay
> Doth grossly close it in, we cannot hear it.

For the Greek scientists, what was even catchier was the idea that each planet was carried around the Earth on a 'sphere' – a thin transparent shell. The shells for the different planets (plus the Sun and Moon) were nested, one inside the other, like the layers of an onion: the Moon was closest to us, and the slowest known planet – Saturn – was the most distant.

There was just one problem: the planets are a bit wayward. Most of the time they move from west to east across the sky, as the months roll on. But at times they can change direction, and travel the opposite way. The early Greeks came up with a complicated fix: each planet was actually transported by four tightly nested spheres that were joined at their poles and rotated in different ways. Phew – heavenly harmony restored!

This theory of the universe was enshrined by the greatest of all Greek scientists around 350 BC. Aristotle, who eventually set up his school in a garden in Athens called the Lyceum, was the world's first serious biologist – though he did occasionally slip up. For instance, Aristotle believed that women have fewer teeth than men: despite being married twice, it seems he never asked either of his wives to open their mouths so he could count their teeth!

He was also fascinated by the heavens. Aristotle polished up the theory of the celestial spheres. To explain the planets' motions properly, he ended up with a total of 55 shells nested together, all moving at different speeds in various directions. If they were singing a refrain, it must have amounted to a cacophony! But a huge revolution was about to shake up these cosy spheres, and reshape the universe. It was all down to one of Aristotle's early, pre-Lyceum pupils, but he was not a scientist, nor a philosopher: he was a soldier – Alexander the Great.

Alexander hailed from Aristotle's home region of Macedonia. But the young firebrand soon turned from studying philosophy to the art of warfare. His ambition was simple: to conquer the world.

By the age of thirty-two, he had succeeded. Alexander's empire stretched from Greece, around the Mediterranean to Egypt, and across the Middle East to India. Alexander intended to conquer China, too, but his homesick soldiers rebelled. And – on his way home – Alexander died.

Alexander overthrew the supercilious view that the Greeks were superior to the 'barbarians' – so-called because the Greeks thought that foreign languages sounded like the bleating of sheep. As he discovered the culture and science of the East, he realised that

– though different – they were every bit as advanced as in his home country. Alexander insisted: 'I do not distinguish among men, as the narrow-minded do … For me every virtuous foreigner is a Greek; and every evil Greek is worse than a Barbarian.'

In this enlightened climate, knowledge flowed freely from country to country. The scientific views of Thales and Pythagoras were cross-fertilised by a stream of accurate measurements of the moving planets that emerged from the newly opened city of Babylon. For over a millennium, Babylonian astronomers had been watching the skies with what we'd now call an astrological agenda. It was imperative to scrutinise the shifting planetary gods, because their motions affected the Earth below. If the planets lined up badly, then the rivers Tigris and Euphrates might flood; the people might rebel against the king; or the harvest might fail.

In the skies above Babylon, the sky-watchers checked out the planets' positions more accurately than anyone before. They began to pick up and record regular patterns and so could predict, from one year's observations, how the planets would behave the following year. In one of these Goal Year Texts we read: 'Nabu [Mercury] is visible with Nergal [Mars] at sunset; there will be rains and floods. When Marduk [Jupiter] appears at the beginning of the year, the crops will prosper.'

These predictions also needed precise calculations, and so the Babylonians invented a new-fangled kind of arithmetic. While we count in 10s (ten, twenty, thirty and so on), the Babylonians chose to count in 60s. They divided a circle into 360 degrees, then split each degree into 60 'minutes', and each minute into 60 'seconds'.

These minutes and seconds were measuring tiny angles, but the same idea was used to divide an hour of time into the 60 minutes – each containing 60 seconds – that tick away on clocks even today.

Strange as it may seem to our eyes, this arithmetic was highly accurate – and the Babylonians really could predict where the planets would lie. For the Greeks, there was just one problem. Aristotle's nested spheres couldn't match the precise data. So what were they to do?

Well, this is the point where history may have to be rewritten – thanks to the discovery of the Antikythera Mechanism.

If we'd been writing this book a few years ago, we'd have said that a very smart astronomer based on the island of Rhodes – Hipparchus by name – came up with the answer. Around 150 BC, Hipparchus drew up the first accurate chart of the 850 brightest stars, and discovered that the Earth's axis is gradually swinging around in space – so the star marking 'north' for Hipparchus was not the same Pole Star we have today. When it came to the planets, Hipparchus threw out the heavenly spheres, and – it's been thought – dreamed up a totally new idea. Historians have lauded it as a supreme example of a purely mathematical achievement – a most impressive example of Greek geekery.

Take the planet Mars, for instance. Hipparchus said that it basically follows a giant circle around the Earth. But the planet isn't actually *on* that circle: it's attached to the rim of a smaller circle – the epicycle – which is carried around by the big circle. You can see this idea in action at a funfair. On a waltzer, you sit in a car that spins around (like Hipparchus's epicycle) all the time that it's speeding around the central pivot. The operator in the centre sees you sometimes travelling forwards, and sometimes backwards – just as we see Mars move through the sky.

The new ideas were picked up later – lock, stock and epicycle – by another hugely influential but cerebral thinker, Claudius Ptolemy. Usually known as the Ptolemaic theory of the planets, this ingenious construction ruled unchallenged for the next 1,400 years – until Nicolaus Copernicus moved the Sun into the centre of the planetary system (and we'll take a look into his thought processes in Chapter 4).

The ticking time bomb that is now threatening to explode this neat view of history came to the surface – quite literally – in 1900. Near the Greek island of Antikythera, divers were plunging down deep to harvest valuable sponges. But they stumbled over something totally priceless ...

'There's a heap of naked women down there!' exclaimed Elias Stadiatos as he surfaced from the first dive. His captain went down to check and returned with the arm of a bronze statue. The divers had chanced upon the wreck of an ancient ship, stuffed full of amazing treasures. There were statues of people and horses, fashioned from bronze and marble, along with rich stores of wine and jewellery. Probably looted by the Romans from the cities of Asia Minor, this rich cargo was on its way back to Italy when the ship went down around 70 BC.

The National Archaeological Museum in Athens restored the best of these treasures and put them on display. But there was one item so odd and unattractive that it lingered behind the scenes: a lump of decaying wood and metal, about the size of a shoebox. The Antikythera Mechanism clearly contained gear wheels, like a clockwork mechanism: yet it dated from over a millennium before the first known clock in the Western world.

'Its discovery was as spectacular as if the opening of Tutankhamen's tomb had revealed the decayed but recognisable parts of an internal combustion engine,' wrote Derek de Solla Price in the 1950s in the British science magazine *Discovery*. A British expert on the history of clocks and scientific instruments, Price travelled to Greece to investigate at first hand. It was indeed a clockwork mechanism, he found, containing at least twenty gearwheels. Unlike a clock, the device didn't run automatically; you had to turn a handle on the side that would then drive the intermeshing gears round at different speeds. Price discovered that the Antikythera Mechanism had a front dial with 365 divisions, along with words for various months and signs of the Zodiac. He guessed that a hand must have moved around this dial, tracing the Sun's motion during the year, while another hand showed the Moon's motion. Price deduced – correctly, as far as he went – that this mechanical device was a celestial calendar.

Arthur C. Clarke was swept away by the importance of the discovery, prompting Price to jokingly request Clarke – who was also passionate about scuba diving – to 'please find some more!' But Clarke didn't think he was up to the challenge of scooping up another important Greek mechanism. Before his death at the age of ninety in 2008, Clarke told us: 'The most advanced underwater artefact I have discovered is an early nineteenth-century soda water bottle!'

'Price was the one to recognise the significance of the mechanism,' says Jo Marchant, who has chronicled the eventful history of this enigmatic device in her book *Decoding the Heavens*, 'and the fact that this revolutionises our understanding of Greek technology.' He couldn't see deeply into the corroded metal lump, though, and had to guess just a little too much. 'He fudged his data,' Marchant continues, 'and was wrong about many of the details of how the device worked.'

But Price's work inspired other equally devoted researchers, right up to the present day. Next to be hooked by the spell was Michael Wright, a specialist in old instruments at London's Science Museum.

'Wright brought a mechanic's eye to this device,' Marchant says. 'He realised Price's errors and made the first accurate working model.' Built from bits of old bronze, Wright's version of the Antikythera Mechanism included hidden gears that he'd located by X-raying the original. The indicator for the moving Moon wasn't just a pointer, he discovered. It was originally tipped with a small ball, white on one side and black on the other, which rotated to show the phases of the Moon.

Even more importantly, Wright realised that the Antikythera Mechanism contained far more gears – arranged in intricate ways – than you needed just to show the positions of the Sun and the Moon. And there were tantalising inscriptions referring to Mercury and Venus. Wright suggested that another set of gears – now lost on the seabed – showed the planets moving around the sky.

The Antikythera spell now ensnared two more British researchers: Mike Edmunds was a Professor of Astronomy at Cardiff University, while Tony Freeth was a mathematician turned television producer. The pair X-rayed the device yet again – but this time with much higher precision. They persuaded the Athens museum staff to look for more fragments in the store-room, and a missing portion from the back was found. And they used new computer techniques to highlight far more of the worn text scratched onto the bronze surfaces.

'Freeth marshalled an international multimillion-dollar team to study the mechanism,' says Marchant, 'and filled in the remaining details that Wright had missed.' They were particularly intrigued by a piece of the mechanism that Michael Wright had discovered, but hadn't been able to explain.

One of the gearwheels that operated the moving Moon had a protruding pin that must have moved along a slot in the adjacent gearwheel. Freeth's team realised that this arrangement made the Moon-pointer move at an irregular pace – exactly matching the Moon's actual motion in the sky. The pointers could reveal precisely the magical moments when Earth, Moon and Sun were in line – and we would be treated to an eclipse.

'The way that this works is completely unexpected, extraordinarily clever and a stroke of genius,' enthuses Freeth. 'It is simple and beautifully elegant.'

With Alexander Jones – an expert on ancient astronomy at New York University – Freeth deciphered many more of the inscriptions on the device. They kept finding references to the planets, and came to the same conclusion as Michael Wright – though they worked out a different arrangement of gears – that pointers on the front of the Antikythera Mechanism showed the moving planets.

Freeth and Jones think that a little gemstone on each pointer identified the celestial body it represented.

They suggest that a gold ball represented the Sun, and the bright half of the Moon was silver. The pointer for Mercury carried a small turquoise, while Venus was depicted by a ball of lapis lazuli. Not surprisingly, Mars was shown as a red crystal of onyx. Brilliant Jupiter was white crystal, and dull Saturn a ball of dark obsidian.

Putting together all the discoveries, the Antikythera Mechanism has now largely been decoded. 'Whoever turned the handle on the side of its wooded case became master of the universe,' Marchant dramatically declares, 'winding forwards or backwards to see everything in the sky at any chosen moment.'

On the front you could perceive the position of the Sun and the phases of the Moon, along with the date in the year and the motion of the planets. The back plate of the device has pointers moving around two spiral grooves. The upper spiral represents a calendar lasting nineteen years – that's a very significant period for astronomers, because after nineteen years the phases of the Moon repeat exactly on the same date. And the lower spiral was marked out in steps of 223 months – the period after which eclipses recur.

To archaeologists, the Antikythera Mechanism sprang into their awareness as a machine totally out of its time. But it didn't arrive in a flying saucer from outer space. Attention is now turning to who made it, and where and when.

First of all, the Antikythera Mechanism can't have been a one-off, a single construction by a genius. Its gearing is so complex it must have evolved from simpler machines. And the gear teeth and lettering are so tiny they are hardly legible; again, someone must have built bigger versions which were easier to construct, and then miniaturised them – just as a mechanical pocket watch today has shrunk from the dimensions of a grandfather clock.

It's not too surprising that we haven't found those ancestors of the Antikythera Mechanism. The Greeks were the ultimate recyclers when it came to a valuable metal like bronze. They were constantly melting down their old possessions to make new ones. We can easily imagine that – when astronomical devices were *so* much last year's fashion – a Greek blacksmith would toss dozens of Antikythera Mechanisms into the furnace, and remould the metal into a statue of the latest handsome athlete to win the Olympic Games. The actual Antikythera Mechanism only survived the furnace because it lay safely on the seabed.

In fact, clues to ancient astronomical computers have been with us all along – but they were ignored because classics scholars didn't believe the Greeks were so gifted with their hands.

'Archimedes fastened on a globe the movements of the Moon, the Sun and five wandering stars which made one revolution of the sphere control several movements utterly unlike in slowness and speed,' wrote the Roman author Cicero in the first century BC. Think of the 'globe' as a circular dial, and doesn't this sound uncannily like the Antikythera Mechanism?

Archimedes lived in the Greek city of Syracuse, in Sicily, around 250 BC. Today, most people remember him as a woolly-headed philosopher, thanks to an incident in the bath. When Archimedes discovered why things float in water, he leaped out of the water and ran down the street crying 'Eureka!' ('I have found it!') – having forgotten to put on his clothes first ...

But Archimedes was also a brilliant engineer. He worked out the theory of levers, showing how you could apply a small force to shift a heavy load – 'Give me but a place to stand,' he enthused, 'and I can move the world!' He devised machines to protect his city from Roman attacks, and designed an angled screw that would draw up water from a well when you turned a handle. Most intriguing, Archimedes used gear wheels – for instance in a windlass, which wound a rope over a barrel to lift heavy weights.

And Tony Freeth sees the hand of the great Archimedes in the history of the Antikythera Mechanism: 'I think the original versions of this sort of device were probably made by Archimedes. The Antikythera Mechanism was made shortly after the death of Archimedes, so it was not one of the machines that he made himself – though it might well have taken much of its design from his machines.'

A key piece of evidence implicating Archimedes is that the months on the Antikythera Mechanism are given names that were used in only a few of the Greek regions, in particular, Sicily. And Freeth has analysed the two calendar spirals on the back, to show that the date-sequence most probably started in 205 BC. That's a few decades earlier than scholars had previously thought; and just seven years after Archimedes was killed by a Roman soldier.

When Archimedes had been born, Aristotle's rotating spheres were all the rage. But the master craftsman would have had problems with this concept. Not only didn't they accurately track the planets, according to the recently discovered Babylonian observations, but Archimedes couldn't easily convert these hypothetical nested shells into a working mechanism. His fertile mind perhaps led him to devise a system of circling gears that made the planets sometimes appear to move backwards, in an early prototype of the Antikythera Mechanism.

And this is where we may need to rewrite history. When Hipparchus came up with his sudden breakthrough of the real planets travelling on epicycles around the Earth, maybe it wasn't such a great intellectual 'aha!' moment as historians have believed. Perhaps he was just studying the intricate gearing of a machine like the Antikythera Mechanism, which had – according to Freeth – been constructed half a century earlier.

'I love this idea,' says Jo Marchant, 'although of course it is very difficult to prove either way. I am increasingly convinced that – rather than devices like the Antikythera Mechanism simply modelling astronomical theories – it was more of a two-way discussion, with physical models inspiring and informing astronomical theories, and vice versa.'

Even after years of investigation, though, the Antikythera Mechanism is full of teasing questions. For instance, who would have commissioned this amazing piece of kit – surely too complex and expensive to be just a toy? Well, we know one set of people who are interested in the exact position of the planets, even today, for their own hocus-pocus predictions. Yes, we're talking about astrology. It would be a true irony if the ancestor of today's computers was built not for science, but rather as an astrological calculator …

Tony Freeth doesn't buy into this theory. 'We now know of between 3,000 and 4,000 Greek text characters on the Mechanism. None of the text that has been read relates in any way to astrology – it's all pure science.'

Marchant suspects the Mechanism was constructed for religious or philosophical reasons. As well as commenting on the device made by Archimedes – before his own time – Cicero describes a similar device that he probably handled for himself. He carries on: 'Suppose someone carried this to Scythia or to Britain. Surely no one in those barbaric regions would doubt that the orrery had been constructed by a rational process?'

'It would have been seen as a statement of everything that was known about the cosmos at that moment in time,' opines Marchant, 'and possibly even evidence that – just as the machine had a creator – so must the universe.'

The best way we're likely to discover more about this mind-boggling invention is to unearth some of the missing parts from the ooze on the seafloor, or to stumble over a sister mechanism there. Starting in 2014, a new team is diving to the wreck of the ancient ship off Antikythera. Jo Marchant hopes 'we might find other similar mechanisms (or other unimagined technology) at this wreck site'. But Tony Freeth is more sceptical. 'In my view, the chances of the *Return to Antikythera* project finding the missing pieces of the Mechanism are zero. It's just remotely possible that they might find another device that was on the same ship: it's probably the best place to look for one.'

This time, the archaeologists are using the world's most sophisticated pressure suit. It looks rather like a spacesuit, but is designed to protect its occupant not from the vacuum of space but from the intense pressure of the deeps – the wreck lies some fifty-five metres below the surface. With its science-fiction appearance and its leading role in underwater exploration, the

Antikythera pressure suit would surely have enthused the Mechanism's lifetime devotee, Arthur C. Clarke.

When we last had the chance to chat with Clarke about this amazing device, he declared: 'If the insights of the Greeks had matched their ingenuity, the Industrial Revolution might have begun a thousand years before Columbus. By this time, we would not merely be pottering around the Moon, we would have reached the nearer stars.'

Chapter 3

ROMANCING THE CALENDAR

Thirty days hath September,
April, June and November.
All the rest hath thirty-one
Except February alone.
That has twenty-eight days clear,
And twenty-nine each leap year.

Cleopatra certainly has a lot to answer for, even after all these centuries. We don't mean just the tacky nightclubs and goods that bear her name – think of our disorderly months, as summed up in the rhyme above. To be sure, the Egyptian Queen didn't write this doggerel, but she did have a hand in our sometimes confusing calendar.

She's also linked to the two months that commemorate real people. July celebrates her lover, Julius Caesar – who created the new calendar – while August was named after her archenemy and nemesis, the Roman Emperor Augustus.

Day in, day out, we live with the calendar that Caesar and Cleopatra have bequeathed to us. But there are other, simpler, ways of counting the dates. We could have a more sensible system from early Egypt; or from the French Revolution; or by adopting the official calendar of Iran.

But we are at the mercy of history – and of astronomy.

The heavens don't make things simple for us. Our lives are governed by the Sun and the Moon – by light and dark; warmth and cold. The Sun rises and sets every day, of course. And as it crosses the sky, the Sun rises to different heights during the year, bringing us the torrid heat of summer and the frigid cold of winter. It would be great if there were an exact number of days in the year. Then our calendar would simply repeat from one year to the next. But nature has to be more complicated. To circle the Sun once, and return to the start of the year, takes the Earth 365 days 5 hours and 49 minutes.

At night, people relied on the Moon for light (before the days of artificial lighting, of course). Even now, if you live away from streetlights, it's amazing the difference the Moon makes – with a full Moon, you can easily walk the country lanes without a torch. The Moon's light changes as it goes around the Earth, which gave ancient people a natural period of time they called a 'moonth' – now our word 'month'.

But the Moon isn't very obliging either. We often think of a lunar month as being four weeks (twenty-eight days), but in fact the period from one full Moon to the next is twenty-nine days, twelve hours and forty-four minutes. So there aren't an exact number of days in the month, nor in the year. And you can't fit an exact number of months into the year, either. How can we even hope to match up the day, month and year to make a sensible calendar?

The first person – in the West – to make a serious attempt was the great Julius Caesar. When he returned to Rome in 46 BC after his conquests in Egypt, Caesar found the Roman state in a mess. With characteristic energy, he tackled everything from improving the welfare system to draining the malaria-ridden marshes around the city. And the existing Roman calendar was among the most hopeless items Caesar had inherited.

Legend told that Romulus – the supposed founder of the city of Rome – had originally divided up the year. Romulus and his twin brother Remus were abandoned in infancy, suckled by a she-wolf and fed by a woodpecker. When the twins argued about the site of their new city, Romulus killed his brother. Turning to astronomy, Romulus decreed that each new year would begin with the month when day and night were equal: he called it March, after his father, the god Mars. There were ten months in all, with the final month hosting Midwinter's Day. That's why December – which for us is the twelfth month – is named after the Latin word for 'ten'.

This ancient calendar left a gap of around sixty days between December and March, which weren't formed into months. A busy city-builder like the mythical Romulus wasn't likely to have slacked off in the winter months. In reality, this calendar was devised by the original farmers of central Italy. After the harvest, the wine-making and planting crops for next year, they didn't need to keep track of time until the Sun's warmth grew stronger and the spring crops began to sprout.

In a big city, though, people had to be controlled all year round. Around 700 BC, Rome's second king – slightly more rooted in reality than Romulus – filled the gap with two new months. Numa Pompilius named

January after Janus, the god who faces both ways, as it marks the re-emergence of the Sun from its winter decline. Falling just before March, newly created February was the last month of the year: it was named after the festival of purification – Februa – held at the full Moon this month. As the last month of the year to be fitted in, poor February was left with fewer days than the other months – a sad fate that hasn't been rectified to this day.

Numa was a priest as well as king, and he appointed himself Pontifex Maximus: literally, 'greatest bridge-builder', but meaning effectively High Priest. And he consolidated his power by keeping the calendar secret, known only to his cronies in the College of Pontiffs. Today, when we can consult a calendar, with its public holidays and important events, at a click of our smart-phone or tablet, it seems unbelievable. Writing in later Roman times, the great philosopher and lawyer Cicero wrote: ' … few people knew whether or not an action could be brought at law at any specified time, for the calendar was not in the hands of the common people. Those who were consulted [the Pontiffs] were in great power.'

In 304 BC, a whistle-blower named Gnaeus Flavius breached the security services of his day. He published the secret calendar on a tablet – made of stone, in his case – which he hung up in the Forum where everyone could see it.

Even so, the Pontifex Maximus had one supreme power over the calendar. The twelve months of the Roman year added up to only 355 days, rather than 365. Every three years or so, this ten-day shortfall had to be made good by adding an extra month, near the end of February. And the Pontifex was the man in charge. Strictly speaking, the Pontifex should have been keeping an eye on the sky, adjusting the calendar when the astronomical signs showed the seasons were getting out of kilter. Like politicians throughout history, though, many a Pontifex used this power to his own advantage. It was easy to add an extra month when your mates were in power, and to refuse to do so to cut short the term of your opponents. Practicalities, like farming, came way down the list of priorities.

Julius Caesar had become Pontifex long before his visit to Egypt, but he had been too preoccupied with waging war to match the calendar with the sky. After his return in 46 BC, the calendar dates had drifted about eighty days from the seasons. People were expected to celebrate the March spring festival while the country was still in the grip of midwinter weather. Caesar's first task was to line up the calendar with the Sun and the seasons, and the year 46 BC took the immediate brunt. The Pontifex added the traditional extra month at the end of February, and then created an extra two months in the autumn, making that year a record 445 days long. On 1 January 45 BC – by this time January had become recognised as the first month of the year – Caesar put the calendar back in synchrony with the sky.

But Caesar wanted a more permanent solution. His methodical mind was now intent on creating a calendar that didn't need constant fiddling from year to year. Never again would there be a situation where – as the later Roman historian Suetonius noted – 'the harvest festivals did not come in the summer; nor those of the vintage in autumn.'

That's where Cleopatra enters the picture. The wily Egyptian Queen was descended from the Greeks who had conquered Egypt with Alexander the Great. Unusually, she took an active interest in the country she ruled – for example, speaking Egyptian as well as Greek. And she lived in Alexandria, the great city on the Mediterranean coast where Greek science was thriving.

Thanks to the young lady's charms, Julius Caesar had tarried in Egypt for longer than intended – including a 'honeymoon' spent cruising the River Nile – and he'd learned a lot about the country's calendar. So he now turned to an astronomer from Alexandria, Sosigenes, to help overthrow the old Roman system and bring in a new calendar: it would go on to take over the world.

The history books don't tell us much about Sosigenes, though legend says he was Cleopatra's personal astronomer. Sosigenes certainly wrote three treatises on astronomy. Under his inspiration, Julius Caesar even wrote a book on astronomy himself, called *De Astris* ('Of the Stars'). The poet Ovid was cynical about this interest: aware that Caesar was beginning to regard himself as a god, Ovid declared Caesar 'wanted to know in advance the heaven that had been promised to him'.

The Egyptian calendar was centred on the Nile: not only was the great river the life-blood of parched Egypt, but it had its own slow and regular pulse. As the Greek geographer and historian Herodotus put it in the fifth century BC: 'The Nile, when it overflows, floods not only the Delta, but also the tracts of country on both sides of the stream, in some places reaching to the extent of two days' journey from its banks.' The flood spread fertile soil across the Nile valley, and then retreated, allowing farmers to raise their crops. The Egyptians chose to begin their year at the life-affirming moment when the Nile flood began, and the start of their year had its own sign in the sky: the first appearance of the brilliant star Sirius, rising in the dawn before the Sun.

And – compared to other calendars – the Egyptian version was elegance itself. They simply counted the days from one rising of Sirius to the next, organising them for convenience into twelve months of thirty days each.

The only glitch in the plan is that the numbered days added up to 360, which was five days short of a whole year. That, they reasoned, was down to the bickering of the gods. Ra, the Sun-god, was angry with the sky-goddess Nut, and forbade her from having children on any day of the year – which was a problem, as she was already heavily pregnant. To help out, Thoth – the god of wisdom – played and won five games of draughts against the other gods: he was awarded five extra days, which he presented to Nut. On each of these days, she gave birth to a child, the new generation of gods. The Egyptians added these five days to the end of the year, so that – in the words of Herodotus – 'the circle of their seasons is complete and comes round to the same point whence it set out'.

The extra days were 'out of time'; and they were considered unlucky. Most people played safe by wearing amulets and taking time off work. Real life started up again immediately afterwards, with the beginning of the New Year. (It's a bit like the period from Christmas to New Year in our twenty-first-century calendar!)

Twelve equal months of thirty days, followed by five days' holiday. What calendar could be simpler? Indeed, the idea was revived after the French Revolution, when – in the Age of Reason – the new government tried to rationalise all the old measures. As well as bringing in decimal currency and the metric system for lengths and weights, in 1793 the French threw out the existing calendar in favour of the old Egyptian system of twelve months, of thirty days each, plus an extra five days. The French called this five-day public holiday at the end of the year *sans-culottides*, or the 'trouser-days', to celebrate that the common people wore trousers rather than the culottes (knee-breeches) of the deposed aristocrats. But the new calendar was deeply unpopular, and – after twelve years – it was abolished by Napoleon.

Going back to Roman times, Julius Caesar decided not to go down the Egyptian route of equal months, but to keep the traditional months. They needed a bit of a shake-up, though.

The Moon is our guiding light to the months. A full Moon repeats roughly every 29½ days. That fraction is a bit inconvenient, but the Moon's phases will repeat around the same date if you alternate a month of twenty-nine days with a thirty-day month. Unfortunately, the ancient Romans were highly superstitious, and regarded even numbers as unlucky. So they had alternated twenty-nine-day months with months containing thirty-one days, rather than thirty days. As time passed, the phases of the Moon drifted out-of-sync with the dates, and some months could even boast two full Moons.

Julius Caesar had no truck with superstition. As you may recall from Shakespeare's play, when a soothsayer predicted his forthcoming assassination – 'Beware the Ides of March!' – Caesar dismissed the man cavalierly: 'He is a dreamer; let us leave him.' And he certainly wasn't going to let mumbo-jumbo interfere with his scheme for a sensible calendar. The Pontifex Maximus kept the 31-day months, but added a day to the others to give them 30 days (except for unfortunate February, which still drew the short straw with its traditional

28 days). The 12 months now added up to the magic 365 days – matching the number of days it takes the Sun to complete its annual tour of the heavens.

But Caesar also learned from the Greek-Egyptian astronomers in Alexandria that there was one extra complication. They knew that a year is not *exactly* 365 days, but nearer to 365¼ days in length. If you watch astronomical events repeat year by year – Midwinter's Day, say, or the dawn rising of Sirius – you'll see that they fall one day earlier every four years.

Two centuries before Cleopatra, a council of priests had exhorted her ancestor Ptolemy III Euergetes to fix this problem by adding an extra 'day out of time' every four years, increasing the total for that year to a six-day holiday period. Although the priests promised future glory for the divine Pharaoh as a result – 'all men may know how the former defect in the arrangement of the seasons ... was made perfect by the god Euergetes' – nothing was actually done.

Until, that is, Julius Caesar and Sosigenes picked up the baton. Caesar declared that every fourth year would contain an extra day, which would be added to February, raising its count to a more respectable twenty-nine days. That is essentially the system of leap years we have today, and it's been running continuously since Caesar inaugurated the new system – known as the Julian calendar – on 1 January 45 BC.

Sadly, Caesar didn't see more than fourteen months of his new calendar, as he was murdered on the ominously predicted Ides of March: 15 March 44 BC. Julius Caesar had been born during the fifth Roman month, Quintilis, and the Senate renamed the month July in his honour. During the very first July, a comet appeared in the sky during games held to commemorate Caesar.

Without Caesar's guiding hand, there was a slight hiccup: the College of Pontiffs was adding an extra day every three years, rather than four. The great Julius's successor, Augustus Caesar, sorted out the mess. He also took the chance to name a month after himself, changing the sixth month, Sextilis, to August. This was his favourite month, because it had seen his major military triumphs – including his defeat of Mark Antony and Cleopatra, which led to her suicide at the fangs of an asp.

There's a popular rumour that Augustus 'stole' a day from February to increase the thirty-day month of August to the same length as July, leading to a reshuffle of the lengths of the months that followed. That's just a mistake made by a thirteenth-century monk, Johannes de Sacrobosco. In fact, February has always had twenty-eight days, and Julius Caesar had already given the other months the days they possess today, including thirty-one days assigned to Sextilis, the month that became August.

What this mediaeval monk did get right, though, is that the Julian calendar is not perfect – it goes seriously adrift over the centuries. We know Sacrobosco worked in Paris, but most of the rest of his life is a mystery. His name means John of Holywood, and he may have come from an abbey of that name in Nithsdale, Scotland. In 1235 he proposed to make the calendar more accurate by dropping one leap day every 288 years.

But nothing happened for centuries. And the push came not from astronomers or farmers, but from clerics, worried about the date of Easter. According to the rules, Easter falls on the first full Moon after the spring equinox, which was defined as 21 March. But – by the mid-sixteenth century – the day when daytime and night-time were of equal length had drifted to 11 March. One of the Pope's advisors, Petrus Ciaconus, warned: 'darkly, like a serpent, this ill shall extend itself more and more each day, unless at some time a cure is necessarily applied'.

His advice spurred Pope Gregory XIII into action. Appropriately, the popes were by then called Pontifex Maximus, the same title Julius Caesar had borne when he changed the calendar. The Vatican's chief astronomer, Christopher Clavius, called for proposals from all over Europe. The best idea came from an obscure Italian medical lecturer, by then dead.

Aloysius Lilius had been born in Calabria, in the 'toe' of Italy, and eventually settled in Perugia. An astronomer as well as a professor of medicine, Lilius spent ten years struggling with the maths of how the calendar was slipping, and came up with a neat solution. After his death, his brother Antonius presented his booklet to the Vatican. It took years of discussion, but in 1582 Pope Gregory gave his full backing to Lilius's plan. In the Julian calendar, every century year was a leap year, because they can all be divided by four. Lilius proposed that a century year was only a leap year if you could divide it by 400. Easy to remember; easy to use. The year 1600 would be a leap year, but not 1700, 1800 or 1900. Most of us have been privileged to live through the only century leap year – 2000 – in the whole period from 1700 to 2300.

Lilius was extolled for his breakthrough. We now call this innovation the Gregorian calendar, but at the time it was known as the Lilian calendar. Clavius referred to Lilius as the 'man most entitled to immortality, who was the principal author of such an excellent correction'. The Gregorian calendar is only twenty-six seconds longer per year than the Earth's actual orbit around the Sun, and it will remain accurate for at least three thousand years.

But Pope Gregory had one more pressing concern. He had to put Easter back on track, and this meant deleting ten days from the calendar. Choosing the month of October, when the fewest religious festivals were held, the Pontifex Maximus decreed that Thursday 4 October 1582 would be followed by Friday 15 October.

Catholic countries quickly changed to the new calendar. But the Protestant states of northern Europe were sceptical about yielding to the Pope's decree, seeing it as religious meddling, rather than a serious application of astronomy. For over a century, travellers had to adjust their dates by ten days when crossing country borders.

The calendar dispute hit the headlines again as the year 1700 approached. If you followed the original Julian calendar, you'd add an extra day to February; according to Gregory's new rules, it would not be a leap year. The gap would grow to eleven days.

The leading astronomer Ole Rømer – along with many fellow scientists – was a passionate advocate of the new calendar, even though he lived in Protestant Denmark. Rømer persuaded the Scandinavian countries to switch to the new system. Sweden – like Julius Caesar in his early days – was preoccupied with war, however, and neglected its calendar. In 1712, the Swedes finally brought their dates into line by adding two leap-days – the only time in history there has been a 30 February!

England was lagging way behind, and trying to do things its own way. Back in the time of Queen Elizabeth, her scientific adviser – and leading astrologer – John Dee had suggested a scheme for revising the calendar, describing his commission in verse:

> As Caesar and Sosigenes,
> The vulgar calendar did make,
> So Caesar's peer, our true Empress,
> To Dee this work she did betake.

He summed up the accuracy of his proposed calendar in the words:

> Three hundred years, shall not remove
> The Sun, one day, from this new match:
> Nature, no more shall us reprove
> Her golden time, so ill to watch.

But all to no avail, when the Church of England objected.

A century later, Sir Isaac Newton applied himself to the problem. The scientific whizz-kid of his age came up with a new formula. As in Gregory's calendar, a century year was generally not a leap year – but you would add a leap day in February if you could divide the year by 500 (rather than 400). In addition – looking a long way into the future – when you could divide a year by 5,000, you'd add two extra days.

That didn't catch on either. By the middle of the eighteenth century, scientific pedantry and religious scruples had to bow to the pressure of international trade and commerce. The world had to have a single calendar, and in 1752 Britain came into line with the rest of Europe by dropping eleven days in September. Later, there were stories that the English people rioted, demanding, 'Give us back our eleven days.' But there's no evidence these riots actually happened: the whole idea seems to have sprung from one of William Hogarth's satirical pictures of an election rally in Oxford, where this declaration appears merely as a political slogan rather than a call to arms. In the British colonies of North America, Benjamin Franklin had a much more laid-back opinion: 'It is pleasant for an old man to be able to go to bed on September 2, and not have to get up until September 14.'

The last major power to fall in line was Russia – but not until after the Bolshevik Revolution in 1917. That's why the glorious October Revolution is commemorated on 7 November.

But the Orthodox Church in the east of Europe was still opposed to the decree of a sixteenth-century Catholic Pope. In 1923, the Serbian scientist Milutin Milanković came up with a solution. Today, Milanković is best known for proposing that Ice Ages are due to astronomical effects – changes in Earth's orbit around the Sun, and a wobbling of our planet's axis. We now know that his pioneering ideas were largely correct, up to the period when human activities began to warm the planet. In his 'revised Julian calendar', Milanković devised the rule that century years would have a 29 February if – when divided by 900 – the remainder was either 200 or 600.

Confused? Well, so was the Orthodox Church. About half of the eastern churches – including Milanković's own church in Serbia – had no truck with the revised Julian calendar, and have stuck with Julius Caesar's original version. So, you will find Christmas and Easter celebrated on any number of different days if you tour the Christian churches in the east of Europe and the Middle East.

And – in some Middle Eastern countries – you'll come across people living by an Islamic calendar that's more accurate than any devised by Christian clerics and scientists. Amazingly, it was developed almost a millennium ago. The calendar's creative genius – Omar Khayyam – was a flamboyant polymath who loved wine, women and song. He's best known today as a poet; translated into English during Victorian times, his *Rubaiyat* is brimming over with Eastern promise:

> Here with a Loaf of Bread beneath the Bough,
> A Flask of Wine, a Book of Verse – and Thou
> Beside me singing in the Wilderness –
> And Wilderness is Paradise enow.

We started our exploration of the calendar with a pair of lovers in Egypt; and now we're ending with an incurable Persian romantic!

Based in what's now north-eastern Iran, Khayyam was hugely respected in his own time as a mathematician: some of his ideas on geometry underpin Albert Einstein's theory of relativity. Khayyam was also a skilled astronomer, and measured the length of the year more accurately then anyone before. The Iranians were already proud of their calendar – which dated back way before Julius Caesar – and welcomed the chance to improve it further. In 1079, Khayyam suggested the best way to square the days and months with the year was not a regular sequence of one leap year every four years, but inserting eight leap years during a period of thirty-three years.

This calendar was already 500 years old when the Western world chose the – less accurate – Gregorian calendar. Adhering to Khayyam's scheme of leap years, Iran and Afghanistan today run their affairs by the world's most perfect calendar.

With ever-more precise calendars, we have tamed the passing seasons of the Earth and the sky into an orderly succession of dates. But we can never conquer the underlying flow of time itself – as Khayyam states so eloquently in his best-known stanza:

> The Moving Finger writes; and, having writ,
> Moves on: nor all thy Piety nor Wit,
> Shall lure it back to cancel half a Line,
> Nor all thy Tears wash out a Word of it.

Chapter 4

THE EARTH MOVES

He fomented the greatest revolution in astronomy; yet his personal life is shrouded in mystery.

To Harvard University historian Owen Gingerich: 'He was more or less a two-dimensional cardboard figure until I went to Uppsala [University] and saw his library. Handling his books with his own annotations suddenly gave me a sense of reality – yes, here's a man, here's his library, here are his tracts.'

Yet Gingerich remains frustrated as to what made this extraordinary person what he was: 'But there are so many questions I would like to ask him, like: when you were young, did you have a girlfriend, did you enjoy music, did you play an instrument?'

That man was Nicolaus Copernicus.

He was born in 1473 in Torun, on Poland's River Vistula. The son of a rich merchant, who dealt in copper – as his name suggests – Nicolaus was the youngest of four children. By the age of ten, the youngster had lost his father, but fortunately, an unmarried uncle took on the family. Uncle Lucas Watzenrode was highly ambitious: he was a business manager for the church's estates. And he saw the potential in young Copernicus. He sent him to study at Cracow, the top university of the day.

It was there that Copernicus started to develop his interest in astronomy. Europe, at that time, was in intellectual ferment. The Crusades had brought the considerable knowledge of the Arabic nations to the West, and in particular the teachings of the Greeks, which the Persians had assimilated. Western scholars translated the Greek works into more accessible Latin. In particular, Ptolemy's mighty *Almagest*, the benchmark of the heavens for 1,400 years, was now available for all to read.

Copernicus was unhappy at the intricacies of Ptolemy's interpretation of the movements of the planets – the 'wheels within wheels' concept of a world circling the Earth, while moving forwards and backwards in its own mini-circle, the 'epicycle'.

Gingerich attributes Copernicus's disquiet to his artistic instincts. He wanted the cosmos to be simple, perfect – not riddled with compromises. 'On the great [astronomical] clock in Strasbourg there's an image with a caption which says: "A true image of Nicolaus Copernicus, based on his own self-portrait". I'd like to know: how did you learn to draw? But we just don't know.' Gingerich continues: 'So this man not only has the talents of an artist, but has the aesthetic sense of an artist. And I think what drove him was the aesthetic sense of unifying things.'

Copernicus went on to study church law and medicine at the University of Bologna in Italy, where he lodged with the professor of astronomy. Their conversations must have gone on well into the night, and it's understandable why the young man was reluctant to leave the heady buzz of learning in Italy to return to the more sedate Poland.

But Uncle Lucas had a job lined up for him. Now bishop of the district of Warmia, he was close to the small northern town of Frombork, which boasted an impressive cathedral. The church was in need of a canon – an administrator – and that was the post Copernicus occupied for the rest of his life. By day, he ran the business side of the cathedral; by night, he would ascend a tower in the cathedral's grounds and observe the motions of the planets. These were the days before the telescope: Copernicus used wooden measuring instruments to plot the positions of his heavenly bodies.

As time went on, Copernicus felt a growing conviction that the answer lay in simplicity. Suppose the Earth was a planet, moving in space. He could then explain the looping of the other planets' orbits as a result of Earth's *own* motion – as it overtook them, it made our neighbour-worlds appear to go backwards.

All of his instincts told him that the planets had circular orbits in space. But *what* were they circling?

'In the midst of all dwells the Sun,' wrote Copernicus. 'For what better place could you find for the lamp in this exquisite temple, where it can illuminate everything at the same time?'

It was an astonishing statement. In one sweep, Copernicus had removed the Earth from the centre of the universe, and relegated it to being a mere planet orbiting the Sun. But the visionary canon hadn't realised the full implications of his work. Copernicus simply printed his sensational findings in a six-page leaflet which he circulated privately to friends. Fortunately, word got around. It reached the ears of Georg Joachim Rheticus, a 25-year-old astronomer from Austria.

Rheticus was a forceful personality, to say the least. He was also somewhat colourful: later in life, he would be sentenced to 101 years' exile for a gay relationship with a student.

Rheticus travelled to Poland to convince Copernicus that he must publish his discovery. The elderly canon was flattered by the enthusiasm of the young man, but it still took two years for Copernicus to write up his findings. He did so in a six-volume work called *De Revolutionibus Orbium Coelestium* ('On the Revolution of the Celestial Spheres'), which Rheticus arranged to be published in Nuremberg.

Copernicus was close to death when *De Revolutionibus* was nearing publication. Unfortunately, Rheticus was away when the printing took place, Copernicus was too ill to write the introduction, and the task was carried out by a clergyman named Andreas Osiander – who clearly didn't get the gist of Copernicus's momentous breakthrough. Osiander failed to grasp that Copernicus had discovered that Earth was merely a planet orbiting the Sun. He saw the canon's work as simply a mathematical sleight of hand that allowed the planets to circle at a constant pace.

Copernicus died peacefully, not knowing the repercussions his book would have. Because of Osiander's obscure introduction (which readers naturally thought that Copernicus had written), astronomers weren't quite sure what to make of it. And the religious authorities didn't get it, either. Otherwise, they would have been appalled at their perfect, sacred Earth being demoted to the status of a mere planet being in the thrall of a much greater Sun. That realisation dawned a century later – and the outspoken Galileo was the unfortunate recipient of the Church's venom.

But word of Copernicus's findings began to trickle through Europe. They reached the ears of larger-than-life Danish nobleman Tycho Brahe, who was born in 1546 – just three years after Copernicus had died. In contrast to the reclusive life of Copernicus, Tycho was a fiery character. During his student years, he got involved in a heated debate with one of his colleagues. Historian Owen Gingerich relates: 'They were arguing about who was the better mathematician. It was a Christmas party, and I think they must have been terribly drunk. They went outside to have a duel ...'

The fight ended fairly decisively when his opponent's sword sliced off the bridge of Tycho's nose. Gingerich flinches: 'It must have been a really bloody mess.' Tycho patched up his nose with a bridge – which, true to his noble status, he claimed was made of silver and gold. But alas for posterity, Owen Gingerich recounts: 'When the Czechs exhumed him in 1901, there was a green stain on the skull. So there's some copper involved in this.'

Tycho was not at all convinced about the findings of Copernicus. The Dane was too religious, and too much of a traditionalist, to abandon the teachings of the Church. But he couldn't take on board the complexities of the Greek ideas about planetary motion: they were complicated, and too unwieldy.

His answer lay in meticulous measurement. Inspired by an eclipse in 1560, and by a close approach of the planets Jupiter and Saturn three years later, young Tycho got his hands on a large pair of wooden compasses to make his own measurements as the worlds appeared to close in on each other. All the predictions of the event were wrong. The Greek estimates were out by a month; Copernicus's by several days.

'What's needed,' Tycho concluded, 'is a long-term project with the aim of mapping the heavens, conducted from a single location over a period of several years.'

He got his single location: the island of Hven, now part of Sweden. It rises from the sound between Denmark and the Swedish coast like a flat cake, and is as unspoiled and agricultural today as it was in Tycho's time.

Tycho landed on Hven in 1576, and declared himself lord of the island. He set about building the most phantasmagorical residence ever seen: a multi-storeyed palace that would incorporate an observatory, living quarters for his considerable number of servants and guests, and a laboratory for alchemy in the basement. Tycho called his scientific mansion 'Uraniborg', after Urania – the muse of astronomy.

His extravagant lifestyle was nothing short of sensational – Hollywood, eat your heart out! Most of his staff would drink several quarts of beer a day, and dinners were celebrations of multitudinous courses: soup, goose liver, venison, lamb, chicken paté, sugar cakes ... all of which might appear in a single evening. And Tycho's staff had to sing for their supper: singing, playing the lute, or reciting poetry. Tycho also had a dwarf, named Jeppe, to entertain the company, and he even kept a pet elk. Sadly, one day it drank too much, fell down the stairs – and died.

But Tycho was serious about his astronomical mission. It was lights out in Uraniborg at 8 p.m., although Tycho and his team of observers would stay up all night. His aim was to measure the positions of the thousand brightest stars – and of the planets – to an accuracy never achieved before, and he succeeded. He was helped by the legacy of Copernicus: his most treasured possession was a set of wooden measuring rulers that had belonged to the great man himself.

Yet time on Hven was running out for Tycho. A new king, Christian IV, had just ascended the throne of Denmark. He wanted to assert his power, and be rid of rebellious nobles – and ostentatious Tycho was right in the firing line. Suddenly, his funding from the crown – which had amounted, percentage-wise, to the amount that NASA receives today from the US Federal Budget – dried up. Tycho had to get out.

But where to? The answer was Prague. The Holy Roman Emperor, Rudolph II, was desperate to have an Imperial Mathematician to add to his court of artists, alchemists and magicians. It didn't take Tycho long to get the job. Ironically, his duties were much more concerned with astrology, rather than astronomy – but in the late sixteenth century, there was far more overlap between the two subjects.

There, Tycho would meet his co-conspirator in advancing our understanding of the universe. And the two men could not have been more different. Johannes Kepler was born in 1571 in the southern German city of Weil der Stadt to a pair of parents out of hell. His father, Kepler recalls, was 'vicious, inflexible, and doomed to a bad end'. He eventually left the family to become a mercenary. Kepler's mother didn't fare any better. She was 'thin, swarthy, gossiping, and of a bad disposition'. To be fair to Kepler, he successfully negotiated to get his mother cleared of witchcraft accusations – she dabbled in herbal medicines. If only Kepler had taken some of her cures; he was, by all accounts, thin, sickly and prone to piles throughout his life.

But he was a brilliant and sought-after mathematician. In the city of Graz (now in Austria), the Protestants had set up a school to rival the Catholic academy – and that's where Kepler was installed. It was a lonely life there; Kepler didn't make friends easily, and had time on his hands to mull over the burning issues of the movement of the planets.

De Revolutionibus had been published fifty years previously, and Kepler was convinced that Copernicus's work was not some theoretical trick: the mathematics stood up too well. And he believed in a stationary Sun.

Soon, religious tensions wound up so much that even Graz succumbed to the Catholic faith. Kepler was banned from teaching, so he had even more leisure to contemplate the heavens. Key to his musings was the arrangement of the planets. He wanted to know how, and why, they move in the way they do. And for that, he needed painstakingly accurate measurements. So, he too had to head for Prague – to meet the Great Dane.

The two unlikely men held their rendezvous at Tycho's estate just outside Prague, in February 1600. Although total opposites – boisterous Tycho and withdrawn, nerdy Kepler – the two men were made for each other. Tycho didn't believe in the Copernican system, but he needed Kepler's mathematical genius to interpret his meticulous observations; Kepler needed those observations to make sense of the Solar System. They were a dream team.

The partnership would not last long, though. In October 1601, Tycho attended a banquet given by a Count Rozmberk. He felt a desperate call of nature – but etiquette didn't allow him to leave the table until his host had departed. Tycho maintained that his bladder burst under the strain – and eleven days later, he died. It was a fitting way out for such a flamboyant character – but can we believe it?

'Of course not!' exclaims Owen Gingerich. 'That would have killed him straight away. Instead, he suffered a slow and painful death, probably of some urinary infection.'

Two days later, Kepler replaced Tycho as Imperial Mathematician. With Tycho's impeccable data, he set to work on his model of the Solar System. But whatever he attempted, Mars just wouldn't fit into place. Kepler had promised Tycho, before his death, that he would polish off Mars in a week. In the event, it took him several years. Publishing his ongoing results in a heavy tome entitled *Astronomia Nova*, Kepler pleaded with his diligent readers – now burdened under a mass of calculations – to forgive him. 'If this wearisome method filled you with loathing, it should more properly fill you with some compassion for me, as I have gone through it at least seventy times ...'

Then Kepler had his 'aha' moment. Copernicus had worked out that the Earth's path was not exactly centred on the Sun; our planet was closer to the glowing orb in January. Kepler suddenly twigged that the Earth changes speed as it moves round the Sun.

After two years' more calculations on Mars, Kepler made his second breakthrough. He realised that Tycho's observations of the Red Planet would fit much better if he 'squashed' the orbit of Mars slightly. 'It was as though I were squeezing a German sausage in the middle,' he explained. But – mathematically speaking – squashed German sausages are hardly the easiest things to get to grips with. Kepler then came up with the concept of an ellipse: the shape you see when you view a circle from an angle. And the figures worked.

Mars, the Earth and the other planets were all in elliptical orbits around the Sun. Gone were the perfect circles of the Greeks; they were replaced by egg-shaped paths which meant that a planet travelled faster when it was closest to the Sun, and more slowly when it was farther away.

This was Kepler's masterstroke. He had found the key that unlocked the vault to understanding the planets' motions – and those of all the other bodies in the Solar System. He enshrined his discoveries in his three laws of planetary motion, a set of equations explaining the orbits of the planets.

But what actually *made* the planets move? Kepler's beady eye was fixed on the Sun. His logic was that the planets' speeds of revolution were at its beck and call. Innermost Mercury races around the Sun in eighty-eight days; the then outermost planet Saturn dawdles in an orbit taking nearly thirty years. Owen Gingerich maintains that the conclusion – to Kepler – was inescapable. 'Because Mercury is the one that's going the fastest, then the driving force has to be coming from the centre – from the Sun.'

To find a force that could drive the planets was a difficult task – to say the least. But Kepler had an idea. He had heard of the work of the English doctor William Gilbert, later to become physician to Queen Elizabeth I. Gilbert was fascinated by magnetism, and demonstrated his experiments to the Queen. Gilbert was convinced that the Earth was a giant magnet, which would explain its power to influence compass needles. And – like Kepler – he also believed that the planets orbited the Sun.

Kepler put two and two together. Tycho Brahe had calculated that the Sun could contain over 100 Earths (we now know the figure to be a million), and it had the power to generate the light and heat that we feel. It stood to reason that the Sun was an even greater magnet than the Earth, which could provide the momentous power to drive the planets around.

It was a brilliant inspiration on the part of Kepler – but it was wrong. The real answer would indeed come from England – a half-century later.

The scene changes to the golden-stone manor house in Woolsthorpe, Lincolnshire. On Christmas morning 1642, a young lad came into the world. It was not an auspicious start: his father had died before the boy was born, and the baby was so feeble that he could have been put into a quart pot, as family legend has it. Yet Isaac Newton went on to become one of the most influential scientists of all time. In the words of Alexander Pope:

> Nature, and Nature's laws lay hid in night:
> God said, 'Let Newton be!' and all was light.

Newton's upbringing was hardly enthralling. Although his father had provided well for the family – they owned a flock of 234 sheep, instead of the usual 30 or 40 – Newton was not happy. His mother remarried, and he was dumped with his grandparents, whom he hated.

He found it difficult to relate to people. At school, he preferred the company of girls – making the furniture for their dolls' houses – and there's even a suspicion that he was a closet homosexual, although he is never believed to have had an affair. Newton was obsessive, withdrawn and geeky. He had no skill in communication. Yet he was undeniably brilliant. That brilliance led him to study at Trinity College, Cambridge. The servants at Woolsthorpe were relieved, and 'rejoiced at parting with him, declaring that he was fit for nothing but the "Versity"'.

At Cambridge, the college fellows were shambolic; they weren't interested in teaching, which also suited the students. It meant that they could get their degrees without doing any work. Newton took advantage of the situation, and taught himself. He elected to choose his own curriculum, concentrating on science and mathematics. As a first-year student, he had no problem in getting to grips with Kepler's obscure books, lapping up the concepts of planetary motion and elliptical orbits.

His curiosity covered everything from perpetual motion to the nature of light. And when a problem gripped him, he became totally absorbed. He would hardly sleep, and neglect to eat. Newton frequently left his food on the plate and, as a result, his cat became notoriously fat.

His fascination with numbers led him to invent a branch of mathematics called calculus: the bane of many schoolkids' lives, yet also the backbone of science and engineering today. But, in typical fashion, he didn't publish his invention. Outreach was not his style.

In 1665, Newton's life went into upheaval. The Great Plague, after sweeping through London, had spread to Cambridge. The only way was out, so Newton returned to Woolsthorpe.

Back at home, Newton's mind was still permanently active. He had been musing a lot on Kepler's dilemma about the nature of the force that drove the planets around the Sun. He watched as an apple fell to the ground from a tree in the garden of the manor house. Many years later, his friend, the polymath William Stukeley (of Stonehenge fame) recalled:

> After dinner, the weather being warm, we went into the garden, and drank thea under the shade of some appletrees, only he & myself. Amidst other discourse, he told me, he was just in the same situation, as when formerly, the notion of gravitation came into his mind.
>
> 'Why should that apple always descend perpendicularly to the ground,' he thought to himself. 'The reason is that the Earth draws it ... and the sum of the drawing power must be in the Earth's centre. If matter thus draws matter, it must be in proportion of its quantity. Therefore the apple draws the Earth, as well as the Earth draws the apple.'

Stukeley continued: 'That there is a power like that we here call gravity which extends itself thro' the universe ... and thus by degrees, he began to apply this property to the motion of the Earth, and of the heavenly bodys ... to consider thir distances ... thir periodical revolutions.'

These reminiscences from William Stukeley are nothing short of astonishing. It would be so easy to believe that the story of Newton and the apple is apocryphal. But Stukeley's account proves that Newton *did* witness the fall of the apple which inspired him to devise his theory of gravity. And the apple tree still stands in the grounds of Woolsthorpe Manor. The original tree was blown down in a gale of 1820. But look carefully, and you'll see that the existing tree grows upwards from the end of a decayed horizontal trunk. Radiocarbon dating in 1998 proved that the tree is over 200 years old. Worship it!

Newton – at the age of twenty-five – had laid the foundations for a new revolution in science. Did he realise it? No. Did he publish? No. He went back to the ivory towers of Cambridge, and pursued his researches in obscurity.

But word had got out. The newly formed Royal Society in London contained some colourful scientific figures, including Edmond Halley (of the famous comet). Renowned for 'swearing like a sea-captain', he was a great diplomat, and a boisterous, lovable man. He was also very persuasive, so Halley was the man who got the job of visiting the reclusive Newton in Cambridge.

It's fair to say that Halley was gobsmacked when he heard Newton recount his results. Not only did they account for the elliptical orbits of the planets; they also explained the nature of the force. It was an 'inverse square law', which means that if you move twice as far from the Sun, then the force diminishes to one quarter. And that is how the planets move. Halley finally convinced the reluctant Newton to publish. And that he did, in 1687: three volumes, entitled *Philosophiae Naturalis Principia Mathematica* – now shortened to *Principia*.

It was one of the four most influential treatises on the universe ever published – along with Ptolemy's *Almagest*, Copernicus's *De Revolutionibus*, and Galileo's *Sidereus Nuncius*. But it nearly didn't get into print because of the failure of a previous Royal Society book on the history of fishes! Halley stood by his guns, and paid for the publication of *Principia* himself.

Principia sets out the underpinnings of how the universe works. After describing the laws of force and motion, Newton uses his laws of gravity to calculate the movements of the heavenly bodies which still hold today – except in extreme conditions, when Einstein's general theory of relativity takes over. In the words of William Stukeley, Newton had 'kept the planets from falling upon one another, or dropping all together in one centre, and thus he unfolded the universe. It was the birth of those amazing discoverys, whereby he built philosophy on a solid foundation, to the astonishment of all Europe.'

Principia was the culmination of a revolution in astronomy that had started with Copernicus just two centuries before. From now on, the momentum would never let up.

Chapter 5

FROM THE DUTCH TRUNKE TO THE HUBBLE TELESCOPE

'Galileo and the telescope.' The words fit naturally together, like Henry Ford and the motor car; or Bill Gates and computer software. But, in reality, Galileo didn't invent the telescope. He wasn't even the first astronomer to turn the telescope towards the heavens. He was simply great at PR!

The telescope had its roots before the great Italian scientist was even born. During the reign of England's Queen Mary (elder daughter of King Henry VIII), the English surveyor Leonard Digges constructed – using 'proportional glasses' – a device which 'discovered things far off, read letters, numbered pieces of money with the very coin and superscription thereof, cast by some of his friends of purpose upon downs in open fields'.

What could be a clearer description of a telescope? But there are problems. Even the most powerful telescopes today could not fulfil the further claim that the proportional glasses could reveal 'what hath been done at that instant in private places'! And Digges' son Thomas, who penned the above account, was an astronomer – yet he wrote nothing to suggest he'd turned his father's amazing device to the heavens. In fact, the glass lenses of Tudor times were so poorly made that all they could have shown was a big blur.

The secrets of the cosmos would literally remain obscure until better glass came along. The breakthrough came not from men of science, but from a secretive enclave of craftsmen on an island near Venice.

As any visitor knows, Venice is *the* place to pick up the most exquisite glassware. At the crossroads of the Roman Empire and the exotic culture of the East, the island city inherited the best of both traditions when it came to making glass. In 1291, the Venetians decreed that all the glass-blowers must move to the small island of Murano. Nominally to prevent their furnaces from burning down the vulnerable wooden city, the edict was actually intended to preserve the secrets of the best glass in Europe.

But industrial secrets have a way of leaking out. By the year 1608, 'Venetian glass' was being made in the important trading port of Middelburg, in the Netherlands. With this high-quality glass to hand, optician Hans Lipperhey (confusingly, he sometimes signed his name 'Lippershey') had the chance to perfect the art of making spectacle lenses.

One day, he peered through two lenses – one held in front of the other – and saw that the distant view was magnified. On 25 September 1608, Lipperhey applied for a patent in the words: 'The bearer of this letter declares to have a certain art with which one can see all things very far away as if they were nearby, by means of sights of glasses.'

The telescope was born.

Rumours rapidly spread that the discovery had actually been made by Lipperhey's apprentice – or even by children playing in the street outside. Then, two other Dutchmen claimed they already had a working telescope. Lipperhey's patent application was turned down. Fortunately, he didn't lose out too much financially. He was commissioned to make several of these new instruments to view the Spanish enemy forces. As a contemporary pamphlet put it: 'The said glasses are very useful in sieges and similar occasions, for from a mile or more away one can detect all things as distinctly as if they were very close to us.'

And it must have been a curious Dutch local who first pointed the new invention at the sky. The pamphlet continues, 'And even the stars which ordinarily are invisible to our sight and our eyes, because of their smallness and the weakness of our eyesight, can be seen by means of this instrument.'

The telescopic cat was now firmly out of the bag. Once they knew it was possible, other opticians could easily experiment with lenses and 'reinvent' the telescope. Within a year, the new device had spread across Europe.

In England, we find the new 'Dutch trunke', as it was called, in the hands of Thomas Harriot. A mathematician and explorer, Harriot had sailed to the American colonies on behalf of Sir Walter Raleigh in the 1580s. It was probably Harriot – rather than Raleigh himself – who introduced tobacco to the British Isles, along with the potato.

Harriot turned the new instrument towards the sky. On 26 July 1609 he drew the earliest sketch of the Moon as seen through the telescope. Another aristocratic patron, Sir William Lower, noted that the Moon 'appears like a tart that my cooke made me last weeke; here a vaine of bright stuffe, and there of darke, and so confusedlie all over'.

A few months later, an Italian entrepreneur turned *his* telescope to the Moon. Galileo Galilei intended the new instrument to be literally his passport to fame and fortune.

A professor of mathematics at the University of Padua, Galileo was also a consummate craftsman. When he heard of the telescope, Galileo – typically – refused to buy one. With the exquisite Murano glassworks right on his doorstep, Galileo was soon manufacturing for himself the best telescopes of his age.

But his first thought wasn't science. Galileo invited the city's senators to the top of the bell-tower, and proved his new device could reveal enemy ships two hours before the regular lookouts could spot them. The Doge of Venice promptly offered Galileo twice his previous salary.

Here was the 'fortune' part of the equation; and now it was time to turn to the 'fame' side. Galileo was convinced his telescopic view of the heavens would put his name among the stars. His finely crafted instruments were several notches above the cheap telescopes that were now the latest 'must-have' gadget, selling in huge numbers at market stalls throughout Europe. 'The telescope had been essentially a carnival toy,' says American historian Owen Gingerich. 'Galileo converted it into an astronomical instrument.'

At the time, Europe resounded with arguments about whether the Sun travels around the Earth, or the Earth orbits the Sun. Galileo firmly believed that our world is on the move, as the Polish canon Nicolaus Copernicus had said. But hard proof was lacking. Like Harriot, Galileo saw mountains on the Moon, and also spots on the Sun. But he wasn't just a casual stargazer; Galileo interpreted his results as a scientist. According to the old Greek concept of the Sun-centred universe, everything in the heavens was perfect. The telescope proved that idea was just plain wrong.

The big breakthrough came in January 1610, when Galileo turned his telescope towards the planet Jupiter. He was amazed to find little 'stars' nearby. As he tracked them night after night, Galileo came to the startling conclusion that they had to be moons, circling Jupiter just as our Moon circles the Earth.

Other astronomers observed these moons around the same time, but Galileo was the only one who latched on to their scientific importance. Critics of Copernicus's theory had said that the Earth couldn't be moving through space, or it would leave the Moon behind. Galileo could now counter with the fact that if Jupiter could take its moons along for the ride, then so could the Earth.

Galileo rushed into print with his new findings, in *Sidereus Nuncius* ('The Starry Messenger') – one of the most important books in the history of astronomy. It would indeed ensure his lasting fame. By obsequiously dedicating the short volume to the Grand Duke of Florence, with a tactful piece of astrology – 'It was Jupiter, I say, who at your Highness's birth ... looked down upon your most fortunate birth from that sublime throne' – Galileo ensured he was invited back to his native city. Some twenty years later, when he got into hot water with the Roman Catholic church over his Sun-centred universe, Galileo gave up astronomy. The

Vatican put him under house arrest in Florence where Galileo devoted his final years to solving fundamental problems in physics.

But Galileo left one outstanding puzzle in astronomy: what was the nature of Saturn? Over the years, the distant planet had changed its appearance. Galileo was baffled – he had an idea but wasn't convinced it was correct. So he covered himself by sending out a letter that included an anagram (in Latin) of the words, 'I have observed the furthest planet to be triple.' If he were correct, he had established the fact that he was the first person to record this strange phenomenon; if he were wrong, no one would know.

His recipient was Johannes Kepler in Germany, the great architect of the Solar System. Applying his formidable mathematical mind to the puzzle, Kepler decoded the anagram as 'Hail, twin companionship, children of Mars.' Amazingly, though Galileo's telescope was too weak to show them, we now know that the Red Planet does indeed possess two small moons.

Kepler was also a huge fan of the new invention. He designed a better telescope, with a differently shaped lens as the eyepiece, and all future lens-telescopes – or *refractors* as they became known – followed this blueprint. Ironically, Kepler's own eyesight was so poor that he never made any observations through a telescope.

For the next half-century, astronomers across Europe built ever more powerful refractors. But they faced one huge problem. The big front lens not only focused light from space; it split the light into a rainbow fringe of colours around the stars and planets. The view was better if you used a weaker lens that brought light to a focus over a longer distance.

As a result, telescopes just 'growed and growed'.

In Danzig – now Gdansk in Poland – a rich brewer called Johannes Hevelius built a telescope 150 feet (45.7m) long, slung from a tall tower. With its immense magnifying power, Hevelius drew the first accurate map of the Moon.

Christiaan Huygens, a Dutch astronomer working in France, disposed with the telescope's tube altogether. His 'aerial telescope' consisted of a lens on top of a tower, controlled by a long piece of string. The astronomer stood on the ground holding the eyepiece in his hand, and looked through it towards the distant lens, hoping everything was lined up in the right direction! Cumbersome though his telescopes were, Huygens could spot dark markings on Mars, and worked out that the planet was rotating. And he at last unravelled the mystery of Saturn's fickle appearance highlighted by Galileo: 'it is girdled by *a thin flat ring, nowhere touching*'.

With another elongated telescope, the Italian astronomer Gian Domenico Cassini found that Saturn's ring was split by a dark band, now called the Cassini Division. Cassini also discerned cloud patterns and a dark 'spot' on Jupiter. Astronomers were beginning to see the other planets not just as brilliant 'stars' in the sky, but as worlds like our own Earth.

But there was clearly a limit to the size of these telescopic giants. Decades passed before the impasse was broken – not by a scientist, but by a barrister living in Essex. Chester Moor Hall was intrigued that the human eye doesn't produce colour-fringes around everything we see. The eye contains both a lens and layers of water; and Hall thought that the combination of two different materials might cancel out the troublesome colours.

In fact, he was wrong: the human eye is pretty imperfect, and the image that's thrown on your retina does have colour fringes. That's why an optician can fine-tune your prescription by getting you to view black circles against red and green backgrounds, and asking which is clearer. In real life, the amazing image-processing of our brains sharpens up the images that we perceive.

But Hall's basic idea was spot-on: colour fringes can disappear if you put together two lenses made of different kinds of glass. Traditionally, everyone had made lenses from the kind of glass you put in window frames (known as 'crown glass'). Around 1700, glassworks began to produce glass containing a lot of lead: it's used in 'lead crystal' wine glasses and decanters with carefully cut facets that gleam all the colours of the rainbow.

In 1733 Hall ordered a pair of lenses – one from each type of glass – to be carefully ground and polished to his exact specification. The two commissions ended up in the hands of the same jobbing lens-maker, by the name of George Bass. The experiment worked: Moor Hall could look through the double lens without seeing any false colours. Satisfied, he quietly went back to his legal work …

In the 1750s, the optician John Dollond was unsuccessfully trying to solve the colour problem when he got chatting with the same George Bass. The lens maker mentioned the pair of special lenses he'd made for Chester Moor Hall. The hint was enough. Dollond soon had a double telescope lens that was free of the troublesome colour fringes. The immediate result was a heated argument about Dollond's subsequent patent application, but for astronomers, it was a miracle. They could build refractors that were powerful, yet relatively short. The first fruits, around 1840, were precision telescopes that could pin down a star's position, allowing astronomers to work out its distance.

Combined with the young art of photography, the new refractors provided stunning images of distant objects in the universe. Attached to a spectroscope that scientifically analysed the light, these telescopes revealed what the distant stars and gas clouds were made of.

The age of the big refractor reached its climax in the late nineteenth century, when the United States emerged as a major player on the international stage. In California, the fabulously rich James Lick – who had made his fortune selling pianos in South America – learned he could immortalise his name by bequeathing money to fund the world's most powerful telescope. Though he had little interest in astronomy, that's exactly what Lick did. His telescope had a lens 36 inches (91cm) in diameter. Astronomers observing here are still kept company by the telescope's patron: a plaque on the huge pier supporting the telescope reads, 'Here lies the body of James Lick'.

A decade later, in 1897, the Lick Telescope was put in the shade by an instrument whose lens was just over a metre across (40 inches). The Yerkes Telescope – which still holds the crown as the world's largest refractor – was funded by Charles Yerkes, a Chicago financial shark. Yerkes made his money from the city's elevated railways and electric streetcar system; he later financed much of the new underground railway in London.

The astronomer who cajoled Yerkes into paying for the new telescope was perhaps the most persuasive entrepreneur in astronomical history. His name was George Ellery Hale.

For Hale, size was everything. With a bigger telescope, you could see fainter objects far out in the universe; and you could analyse their light in more detail. Hale's achievement was to build, in succession, the three largest telescopes of their kind the world had ever seen.

Having commissioned the supreme lens-telescope, Hale now sounded the death knell for big refractors. A lens any bigger than a metre across would sag under its own weight as the telescope moved, he calculated, so distorting the astronomer's view of the distant cosmos. Instead, Hale turned to mirrors. A shiny surface curved like a shaving mirror can focus light from the sky – just like a lens – but has the advantage that you can prevent a mirror from sagging by supporting it from behind.

The idea wasn't new. Isaac Newton had built the first reflecting telescope in 1668, though his pioneering 'reflector' was little more than a toy. This design hit the headlines in 1781, when a musician and amateur astronomer named William Herschel turned his home-made reflector on the sky – and stumbled across a new planet beyond Saturn (more on his discoveries in the next chapter).

Herschel went on to build a reflector with a mirror four feet (1.2m) across. Though it was bigger than the lens on the later Yerkes refractor, built a century later, Herschel's giant telescope didn't change the world of astronomy; indeed, he used it relatively little, except to impress visitors. One problem was its sheer unwieldiness: Herschel needed two assistants to manoeuvre the great beast while he was making observations. Its huge wooden frame wouldn't pass today's Health and

Safety laws: there were several accidents, and Herschel was once almost crushed to death when removing its massive mirror. As he grew older, he reverted to his smaller telescopes.

But none of these drawbacks daunted the enthusiasm of an Irish peer whose eyes were firmly set heavenward. The Third Earl of Rosse came into his family inheritance in 1841, and devoted his fortune to building the world's biggest telescope in the middle of the Irish bogs, at Birr in County Offaly. This reflector had a mirror six feet (1.8m) across, and the giant telescope – controlled by a pair of assistants – was slung between two massive walls. After the contemporary name for Birr, the mighty telescope was dubbed 'The Leviathan of Parsonstown'.

The Leviathan gave Rosse more detailed views of fainter objects than anyone had seen before. When he turned it towards a fuzzy patch in the northern sky called Messier 51, he was in for a surprise. 'Spiral convolutions,' he noted. 'With each successive increase of optical power, the structure has become more complicated …'

His sketch shows what looks like a cosmic maelstrom, with spiral 'arms' reaching outwards towards a fainter companion – giving it the nickname 'The Whirlpool'. Over the next few years, Rosse listed fourteen similar beasts in the heavens.

For all his enthusiasm, Rosse's giant telescope was in the wrong place. Thanks to the Irish weather, he spent much more time kicking his heels in frustration than in observing the sky. It also ruined the heart of his telescope, the giant mirror. Since the time of Newton, astronomers had made their mirrors from solid metal – an alloy of copper and tin, with a little arsenic also thrown into the mix. Known as speculum metal, this alloy wasn't particularly good at reflecting light, even when the surface was freshly polished. And a speculum mirror quickly tarnished when exposed to damp air.

Silver would be a much better bet, and Rosse made some small mirrors of this expensive metal. But even a wealthy landowner couldn't afford to make a six-foot mirror of pure silver! Rosse did try to improve his telescopes by depositing – unsuccessfully – a brilliant layer of silver on the front of a speculum mirror.

Unknown to British astronomers, the key to the future was displayed in the Crystal Palace, the vast pavilion built to house London's Great Exhibition of 1851. Once again, the breakthrough came from master craftsmen, designing an eye-catching new range of ornamental glassware. The exhibits included a set of vases that had been chemically coated with a dazzling film of silver.

One contemporary writer sniffily recorded: 'We cannot think that the pure white glass is in any respect improved by silvering.' But two scientists based on the continent – Carl August von Steinheil in Munich and the Parisian Leon Foucault – had other ideas. Both realised it was much easier to shape a telescope mirror out of glass than from speculum metal. By depositing a very thin layer of silvering on the front surface, you'd create a highly reflecting mirror.

As the twentieth century dawned, George Ellery Hale in the United States was being seduced by the new silver-on-glass reflectors. He left Chicago for the clearer skies of California, and built an observatory on Mount Wilson, in the mountain ranges behind Los Angeles, with specialist telescopes for studying the Sun. Hale had set his heights much higher – both astronomically, and in terms of his telescopes. He first started building a reflector with a mirror 1.5 metres across: far bigger than the lens of the Yerkes Telescope, but not as large as the – now moribund – Leviathan in Ireland. This telescope wasn't even complete when his ambition soared even higher: to create a mirror 2.5 metres in diameter, as the heart of a Hundred-Inch Telescope.

What he needed was money – and Hale had an amazing knack of persuading rich people to part with their dollars. After his success with Yerkes in Chicago, he befriended one of the wealthiest men in San Francisco: John Dagget Hooker had made his fortune in steel and, as an amateur astronomer, he was a natural target for Hale.

Hooker agreed to pay the cost of making the world's largest mirror. According to rumours, he wanted to buy himself into Hale's favour, as the astronomer had become a very close friend of Hooker's wife and her best friend, to the exclusion of Hooker himself. In the end, Hooker became so jealous that he forbade Hale from visiting his household. And he never paid the final instalment on the mirror. Hale extracted the rest of the observatory costs from the Carnegie Institution of Washington. Despite the rift with his sponsor, Hale magnanimously named the world-beating instrument the Hooker Telescope.

Opened in 1917, the mighty power of the Hundred-Inch Telescope was matched a couple of years later by the peerless mind of Edwin Hubble. A former lawyer, Hubble turned his analytical mind to weighing up new evidence that the great telescope was uncovering about the distant universe.

Hubble swung the Hooker Telescope towards some of the spiral objects that the Third Earl of Rosse had discovered. The vast size of the mirror meant he could pick out individual stars within the glowing masses, and then measure their distances. Hubble told the astonished world that these spiral objects were actually giant star-systems, like our Milky Way, but lying millions of light years away. It was the first proof that the universe is populated with billions of other galaxies. (Look forward to reading a lot more about Hubble's achievements in Chapter 12.)

Telescope-supremo Hale was convinced there were further discoveries to be made, beyond the reach even of the Hundred-Inch Telescope. His motto was 'Do not make any mean plans'. In 1928, Hale wrote an article in *Harper's* magazine, pushing the case for a larger telescope. It began: 'Like buried treasures, the outposts of the universe have beckoned to the adventurous since immemorial times.'

The oil magnate John Davison Rockefeller read only this first sentence. He phoned up Hale, and ended up pledging six million dollars for a new Two-Hundred-Inch Telescope. With a five-metre mirror, it would be able to see twice as far into space as the Hooker Telescope.

Hale's dream telescope was completed in 1948, ten years after the death of its founding genius: appropriately, it's named the Hale Telescope. Still informally called the Two-Hundred-Inch Telescope, it made the most important observations of the last half of the twentieth century, including the discovery of the stupendously powerful quasars, now thought to harbour gargantuan black holes (see Chapter 11).

More recent telescopes have grown to eight metres in diameter, but it's almost impossible to make – and transport – a mirror any bigger than this. The world's largest telescopes are now constructed from smaller mirrors that are hexagonal in shape and fit together like bathroom tiles to create a single large reflecting surface. The pioneers of these segmented-mirror behemoths were the pair of Keck Telescopes, perched on the summit of Mauna Kea, a mammoth extinct volcano in Hawaii. Situated at almost half the height of Mount Everest, the Keck Telescopes have an unrivalled view of the sky. Despite their altitude, though, these telescopes still have to peer upwards through the Earth's atmosphere, with its turbulent air currents. For the discriminating eye of a giant telescope, it's like the distorted view we see when we look upwards from the bottom of a swimming pool.

The ultimate solution has to be a telescope in space. First conceived way back in 1946, the Large Space Telescope took almost half a century to be completed and launched high into orbit around the Earth. Along the way, it was renamed after America's greatest extragalactic astronomer, as the Hubble Space Telescope.

Astronomers gave it the largest mirror they could fit inside the belly of the Space Shuttle. At 2.4 metres across, the Hubble Telescope is – by pure coincidence – almost exactly the same diameter as the Hooker Telescope used by Edwin Hubble himself to explore the universe in the 1920s.

Though it's far from being the world's largest telescope today, the Hubble Telescope's crystal-clear view of the universe lets it focus on exquisite details. Who can forget its incredible view of the birth of stars, in the vast dark silhouettes of the 'Pillars of Creation'?

The next space telescope will be even wider than the rocket that's launching it. Named after NASA's pioneering administrator, the James Webb Space Telescope has a segmented mirror that's folded up for launch. After it reaches space, the mirror will unfurl like a cosmic flower, into a vast eye on the cosmos that's almost three times the width of Hubble's mirror. We'll have to wait until after 2018 to see the harvest that the James Webb will bring in. But the Hubble Telescope has already scooped a rich bonanza. It has seen farther than any telescope on Earth, out to horizons over 10 billion light years away.

Here, Hubble has become a time machine. The light from these distant galaxies left them aeons ago, to speed through empty space towards us. As a result, Hubble sees them as they were in their infancy, soon after the universe itself was born in the Big Bang. From these 'baby pictures', astronomers can work out how chaotic clouds of gas in the early cosmos have evolved into the stately spiral galaxies we see today, with their majestic stars and homely planets.

Only 400 years ago, Lipperhey and Galileo were amazed to see armies and ships, the Moon and planets, in close-up. How astounded would they be to know of the vast distances in space – and the depths of time – that have been plumbed by the descendants of their humble 'Dutch trunkes'.

Chapter 6

PLANET SEEKERS

'A curious either nebulous star or perhaps a comet,' reads the entry in William Herschel's observing log for Tuesday 13 March 1781.

Quite an understatement for a discovery that would double the size of the Solar System … and revolutionise astronomy.

William Herschel – originally Friederich Wilhelm Herschel from Germany – was the son of a musician in the Hanoverian Foot Guards. Inspired by his father, he took up the violin and oboe as a boy. He would later become an oboist for the Hanoverian band, before he left for England after the defeat of the army by the French at the Battle of Hastenbeck in 1757.

As well as music, Herschel senior also instilled a deep love of astronomy in the young man. William's devoted sister Caroline reminisced how their father inspired them both with his vision of the heavens. 'My father was a great admirer of astronomy; for I remember his taking me out on a clear frosty night into the street, to make me acquainted with several of the beautiful constellations, after we had been gazing at a comet which was then visible.'

Caroline – herself an accomplished astronomer and musician – would later join her beloved brother at his elegant Georgian terraced house at 19 New King Street in Bath. That's where William – having held posts as a performer in the North of England – finally got his coveted job: organist at the fashionable Octagon Chapel in the city. Here he was free to perform – and compose – music.

Caroline looked after the housekeeping, while William pursued his passions. But like many musicians, William was obsessed by mathematics. He devoured a treatise on harmony written by the Cambridge mathematician Dr Robert Smith. A fan of the author, Herschel then followed it with another work by Smith: 'A Compleat System of Opticks'. He became fascinated with the descriptions of telescopes it contained; and soon William was gravitating towards what would prove to be the major love of his life – astronomy.

Herschel's cosmic mission was to plumb the depths of space. While professional astronomers strove to prove Newton right (or wrong) about gravity, the amateur William wanted to measure distances to the stars. He reasoned that bright stars (which he assumed were nearby) should show movement against the distant stars, as seen from the perspective of the Earth orbiting the Sun.

To probe the starry skies, Herschel had to make a thorough survey of the heavens. And for that, he needed a telescope. But how could an impoverished musician afford a telescope? Answer: make one. The easiest type to construct was a reflecting telescope: a giant version of a shaving mirror. And William was determined to make the biggest he could.

In Herschel's day, the mirrors weren't made of glass, but of metal alloy. And William was so obsessed with finishing a mirror that he wouldn't take his hands off it for over fifteen hours. So Caroline fed him by hand as he worked with the molten metal. One day, the metal poured out onto the floor, and the flagstones cracked: ' ... and the caster and his men were obliged to run out at opposite doors,' recalled Caroline, 'for the stone flooring flew about in all directions, as high as the ceiling. My poor brother fell, exhausted with heat and exhaustion, on a heap of brickbats.'

But all this effort was worth it. Herschel was conducting one of his cosmic surveys on the night of 13 March, when he detected his 'nebulous star'. That night, he probed his mystery object with higher and higher powers through his seven-foot reflector, which, as the Astronomer Royal himself would observe later, was 'of extraordinary excellence'. Since the object grew bigger with increasing magnification, Herschel was convinced that it had to be a comet. If it *was* a comet, it would move within a period of days.

Four nights later, Herschel revisited the spot in the company of his friend, the physician Dr William Watson. 'I looked for the Comet or Nebulous Star, and found that it is a comet, for it has changed its place,' wrote Herschel.

Watson was suspicious about the nature of Herschel's object. Could it be something much more significant than a comet? He wanted to alert other astronomers to the discovery and – by great fortune – his father held high status at the prestigious Royal Society in London, home to the most distinguished scientists in the world. Watson urged his friend to send a formal account of his findings to Watson senior. Meanwhile, he acted as go-between, asking his father 'to be so kind as to take proper measures to have this Paper read on Thursday so that Astronomers may have immediate notice of this event'.

The paper had the desired effect: astronomers flocked to their telescopes. But the next few months were wracked with confusion. Herschel's telescopes were designed for high magnification, not for measuring accurate positions. Professional telescopes, on the other hand, were precision measuring-instruments. Some astronomers, with their low magnifications, had difficulty even in identifying Herschel's mystery object. But as time passed, enough observations accumulated to prove that Herschel's 'Comet' was undoubtedly a planet. A planet – incredibly – twice as far out as Saturn, four times larger than the Earth, and almost fifteen times heavier.

Although Herschel was certainly the discoverer of the new planet, he wasn't the first astronomer to spot it. That honour goes to Britain's first Astronomer Royal, John Flamsteed, who marked it as a star – 34 Tauri – on a chart of 1690. There were at least fifteen other sightings before Herschel's discovery, although – to be fair – none of Herschel's predecessors had telescopes with his light-grasp and magnifying power. But there are no excuses for a French astronomer, who recorded his observations in a somewhat disposable fashion – scribbled on a paper bag that had contained hair perfume!

Now astronomers were up against a dilemma they'd never faced before. What to call this new planet? Herschel pushed for 'Georgium Sidus' (George's Star), in honour of George III, the king of his adopted country. But European astronomers were less than enamoured with a planet called George. Led by the highly influential German astronomer Johann Bode (director of the Berlin Observatory), the Europeans pressed for a more traditional mythological name. Working outwards through the Solar System, Mars's father is the next planet, Jupiter; his father was Saturn; and so the planet beyond Saturn should be named for Saturn's father – Uranus, the Greek god representing the sky.

The king's nose was not put out of joint by being passed up for planetary posterity: in fact, George was hugely enthusiastic. Although his mental health declined dramatically in the later years of his long reign, leading to reports that he was 'mad' (he may have suffered from the blood disease porphyria), his younger years were vigorous and productive. Two of his passions were agriculture (he was nicknamed 'Farmer George', on account of his fascination with the land) – and science. In fact, George III is the only British monarch known to have been interested in science. He had his own laboratory, and even a small observatory at Kew. He's quoted as saying: 'I spend money on war because it is necessary, but to spend it on science – that is pleasant to me. This object costs no tears; it is an honour to humanity.'

And now he was about to confer the greatest honour on William Herschel. He wanted William to become his private astronomer.

William Watson Senior, that invaluable insider at the Royal Society, wrote to Herschel on hearing of the king's decision: 'My dear friend. I need not tell you with what satisfaction I received the account of the King's offer to make you independent of music. You are now upon the eve of entering into a new course of life … to be in a situation which will at the same time command respect and, what is still more desirable, enable you wholly to give up yourself to an employment which is attended with the highest gratification.'

On hearing that the salary was to be fixed at £200, Watson went on to comment: 'Never before was honour purchased by a monarch at so cheap a rate.'

For Herschel, there was no holding back. His mind was totally focused on astronomy – not music. Caroline noted, 'the prospect of entering in on the toils of teaching, etc … . appeared to him an intolerable waste of time'. The deal they struck was – as Caroline recalled – 'I should give up my music profession, and, settling somewhere in the neighbourhood of Windsor, devote my time to astronomy'.

William and Caroline eventually settled in then-rural Slough, close to Windsor Castle. There he was free to build ever-more gargantuan telescopes, and continue his beloved survey of the heavens. His greatest construction was the mighty 'Forty-Foot', which boasted a mirror four feet (1.2m) across! No less a figure than Joseph Banks, President of the Royal Society, added his congratulations: 'My best compliments to Mr. Herschel with wishes that for the sake of science his nights may be as sleepless as he can wish them himself.' The great and good came from all over the world to meet the remarkable man whom the king had installed. And he never seemed to tire of showing them the wonders of the universe through his colossal telescopes.

Meanwhile, Caroline, as well as helping her brother and singing his compositions, was busy with observations of her own. Inspired by the sight of the comet that her father had shown her in the sky when she was a girl, she went on to discover eight for herself – the first woman to achieve such a feat. She also embarked on correcting the star catalogue that had been produced by John Flamsteed in the early eighteenth century. And she went on to compile a list of 2,500 nebulae – mysterious fuzzy patches in the sky.

In 1828, the Royal Astronomical Society in London awarded her its prestigious gold medal. It would not be until 1996 that the honour would be bestowed on a woman again – this time to Vera Rubin, a pioneering researcher into galaxies.

Fast-forward to 3 July 1841. By now, Uranus had completed nearly three-quarters of its 84-year orbit around the Sun. And it wasn't behaving itself.

A young Cambridge undergraduate, John Couch Adams, wrote in his diary: 'Formed a design at the beginning of the week, of investigating, as soon as possible, the irregularities in the motion of Uranus which are yet unaccounted for; in order to find whether they may be attributed to the action of an undiscovered planet beyond it, and if possible, thence to determine approximately the elements of its orbit etc., which would probably lead to its discovery.'

John Couch Adams was born on 5 June 1819 at Lidcot – a farmhouse in Laneast, Cornwall. His father, Thomas, was a tenant farmer; his mother, Tabitha, a devoted wife. Poor but happy, they gave John an idyllic upbringing which he never forgot. But schooling the boy in the wilds of rural England proved a problem. John was forever outpacing his teachers, and proving to be an excellent mathematician – a subject in which he was largely self-taught.

He was also developing an interest in astronomy. On 17 October 1835 he observed Halley's Comet, and described it to his parents: 'You may conceive with what pleasure I viewed this, the first Comet I ever had a sight of, which at its visit 380 years ago threw all Europe into consternation, but now which affords the highest pleasure to astronomers for proving the accuracy of their calculations and predictions.'

By now, it was clear that John must go to university. Fortunately, his family had come into a small legacy, which would help with the heavy college expenses. And in the spring of 1839, the young man began studies with the Reverend George Martin, a Cambridge graduate and good mathematician. It was probably because of Martin's influence that Adams decided to apply to St John's College, Cambridge. By the autumn of that year, he was ready to leave home.

Adams had a gentle and kind nature, winning friends at the university quickly. They were also soon to realise his proficiency at mathematics. Fellow student A.S. Campbell recalled meeting Adams, and feeling a mixture of admiration and despair: 'I had gone up to Cambridge with high hopes, and now the first man I meet is something infinitely beyond me … if there were many like him, my hopes of success were gone. A few days' experiences soon relieved me and I knew what a wonderful man I had met. If I could keep *near* him in the examinations I should do very well.'

Tripos – the ultimate maths examination at Cambridge – took place for both Adams and Campbell in January 1843. This marathon of an exam, comprising eighteen three-hour papers, was the acid test of a candidate's scientific ability. There was no learning by rote: students had to think on their feet. Those men who qualified for the highest class of Tripos honours were called Wranglers, ranked numerically in order of performance – from the Senior Wrangler down.

Campbell, by now Adams's close friend, watched John sitting Tripos. 'I noticed that when everyone was writing hard, Adams spent the first hour in looking over the questions, scarcely putting pen to paper the while. After that he wrote out rapidly the problems he had already solved in his head and ended by practically flooring the papers.'

Not to anyone's surprise (except, perhaps, to Adams himself), the brilliant young man came out as Senior Wrangler. He scored more than 4,000 marks in an examination in which the Second Wrangler managed less than 2,000. (Campbell himself came out as Fourth Wrangler – no doubt to his great relief! He became a Fellow of St John's.)

After his graduation, Adams was kept busy with teaching duties at St John's. But he never forgot the promise to himself of 1841, and resolved to return to settle the question of 'wobbly' Uranus.

Now three new major players enter the scene: James Challis, Director of the Cambridge Observatory; George Airy, Astronomer Royal at Greenwich; and Urbain Le Verrier, of the Paris Observatory. And here's where the waters start to go murky.

Scholars of the highest calibre have scrutinised the events of the next few years. No overall vision emerges – because here we enter the world of clashing human personalities, missed opportunities, meetings that never happened, misinterpretations of data and the approach to that data, buried grudges ... and apocryphal stories galore.

In short, what happened next was a farce.

Adams believed that he was close to solving the problem, but needed data from the Astronomer Royal at Greenwich. He quite correctly made his request through James Challis, Director of the Cambridge Observatory: 'I applied to Mr Airy, through the kind intervention of Professor Challis, for the observations of some years in which the agreement appeared least satisfactory. The Astronomer Royal, in the kindest possible manner, sent me, in February 1844, the results of all the Greenwich observations of Uranus.'

By 1845, Adams felt that he knew where the missing planet was.

Now was the time to send his results to George Airy. As he was about to leave Cambridge to visit his parents' home in Cornwall, he came up with the idea of delivering his results to Airy at Greenwich en route. He prevailed on Challis again, who wrote a helpful letter of introduction: 'My friend Mr. Adams (who will probably deliver this note to you) has completed his calculations respecting the perturbations of the orbit of Uranus by a supposed ulterior planet, and has arrived at results which he would be glad to communicate with you personally, if you could spare a few moments of your valuable time.'

Before leaving for Greenwich, Adams left a summary of his data on the postulated new planet with Challis. As well as the usual information – how distant the planet lay from the Sun, and how heavy it was – he added a big hint: *where* the planet should appear in the sky on 30 September 1845. Adams had hoped that Challis might search for the undiscovered world with his Northumberland Telescope, which – with a lens 11.6 inches (29.5cm) across – should have been capable of seeing the planet.

Alas, Challis did not take the hint. Although a former Senior Wrangler himself, he was nowhere close to Adams in his mastery of science. A colleague once described him as 'courteous in manner, kindly in disposition, simple and unassuming in character'. It seems that Challis never quite wore the trousers. He was married to a small and feisty lady, who is rumoured one night to have pulled a large burglar from under the couple's bed, while Challis cowered in the next room. And one night, she was worried about her husband observing so late. She rushed into the Northumberland dome – to find Challis trapped behind the telescope.

Meanwhile, in Greenwich, Adams was having no luck in intercepting the Astronomer Royal. On one occasion, Airy was in Paris. Next time, he was not at home, but at least Adams was able to deliver his card and his results. Later that day, the young man tried again, but was told by Airy's butler that on no account should the Astronomer Royal be disturbed, because he was having his dinner.

With his tail between his legs, Adams returned to Cambridge feeling hurt and distinctly rebuffed.

Why did Airy ignore Adams? To understand the nature of the man, you have to go back to his childhood. Airy was born in the north of England, and spent most of his time as an unsocial boy, obsessed with schoolwork. At the age of twelve, Airy learned double-entry bookkeeping – a system that would govern his approach to practicality and order for the rest of his life. As one of his colleagues at Cambridge later remarked: 'His nature was eminently practical. And his dislike of mere theoretical problems was proportionately great. He was continually at war with some of the resident Cambridge mathematicians on this subject.'

As Astronomer Royal, he had to push the envelope way beyond astronomy. With the Industrial Revolution in full swing, he turned his mind to matters of practical importance, such as the width of the track gauge on the Great Western Railway and the cause of the Tay Bridge rail disaster. Effectively, he was also Chief Government Scientist. Airy was acutely aware of his responsibilities. And he was mightily unwilling to sacrifice his governmental duties for chasing up what might be a flash-in-the pan idea.

As well as being obsessed by practicality and order, Airy was also an extreme perfectionist. For him, people came in two groups: those who had succeeded, and were worthy of cultivation; and those who hadn't, and weren't. Young people almost inevitably fell into the latter category. So it's hardly surprising that Adams, aged twenty-six, failed to impress. The reception to Adams's hypothesis at the Royal Observatory was cool. And Airy himself – although he tried to conceal it – had a strong negative reaction to the paper.

Meanwhile, across the English Channel, things were stirring on the cosmic front. The French mathematician Urbain Le Verrier was on the same track as Adams, and was savvy enough to publish his predictions. But his obsession with efficiency made him so disliked in France that none of his colleagues wanted to help in the search for the mystery world. As one of his colleagues remarked: 'I do not know whether Monsieur Le Verrier is actually the most detestable man in France, but I am quite certain that he is the most detested.'

However, Airy did not detest Le Verrier. In June 1846, he received the Frenchman's 'memoir' on the missing planet. 'I cannot sufficiently express the feeling of delight and satisfaction which I received from it,' he wrote.

Le Verrier's predicted position for the new planet agreed with that of Adams to just one degree. Yet Airy still held doubts about the younger man's work. But he called in Challis to make a search – reasoning that Challis's telescope, the Northumberland, was more powerful than Airy's Sheepshanks refractor.

Challis embarked on a reluctant hunt for the missing planet. His confidence in Adams had been dented by Airy's rejection of the young man; and it didn't help that Challis was hardly on Airy's A-list when it came to his own astronomical talents. On one occasion, Challis *did* find an object with a disc. It never seemed to occur to him to observe it at a higher power. Next night, a guest of Challis at a Trinity College dinner suggested that he should – so Challis invited his colleague to come to the observatory and check the object out. When the two men arrived, the skies were clear; but – alas – Mrs Challis insisted on giving them both a cup of tea before they commenced their observations. By the time they got to the Northumberland Telescope, it had clouded over!

Meanwhile, back in France Le Verrier was becoming increasingly agitated. Although he knew that Challis was conducting a search, none of his French colleagues would collaborate with him. Finally, in mid-September 1846, he ran out of patience.

Le Verrier was in the pits as to where to turn. And then he remembered. About a year before, he had received a doctoral dissertation from an assistant at the Berlin Observatory, Johann Gottfried Galle. Either Le Verrier had failed to respond, or the correspondence had been lost. Either way, Le Verrier remembered the young man's talent – and poured out an effusive letter to him:

> Sir – I have read with much rigour and attention the reduction of Rømer's observations which you have been kind enough to send me. The perfect clarity of your explanations ... are on a par with those which we should expect from a most able astronomer.
>
> Right now I would like to find a persistent observer, who would be willing to devote some time to an examination of a part of the sky in which there may be a planet to discover. I have been led to this conclusion by the theory of Uranus ...

With the letter, Le Verrier included his predictions for the position of the planet, its orbit, and even its expected size as seen through a telescope of 'good glass'. It also included a greeting to Johann Encke, Director of the Berlin Observatory, whom Le Verrier had never met.

When the letter arrived five days later, on 23 September 1846, the enthusiastic young Galle tried to persuade his superior that he should search. But Encke's attitude to young people was on a par with that of Airy: dismissive, to say the least. Galle persisted, and in the course of his discussions with Encke, the two men were interrupted by Heinrich Louis D'Arrest, a young student astronomer. He begged to be involved in the search.

That night, Galle and D'Arrest headed for the pride of the Berlin Observatory – a nine-inch (23cm) refractor with a lens crafted by Joseph Fraunhofer, the leading glassmaker of his day who had died twenty years earlier. Galle searched Le Verrier's target area – looking for a disc – but came up with nothing.

D'Arrest, hovering in the wings, suggested a more accurate star map than Galle was using. D'Arrest leafed his way through some disordered charts, and eventually came up with one that had been published as recently as 1845. The Hora XXI of the Berlin Observatory's Star Atlas had hardly been used – and had not yet been distributed to other observatories.

The men resumed their work – Galle at the telescope, D'Arrest sitting at a desk, checking the atlas, while Galle called out the positions and appearance of stars that he could see through the telescope. Very shortly into the new regime, D'Arrest gasped. 'That star is not on the map!' he exclaimed.

On 23 September 1846, our Solar System acquired a new planet.

The new kid on the block was virtually a twin of Uranus. It was made of water and gas, just slightly smaller than its recently discovered cousin, but heavier – it would outweigh the Earth seventeen times.

Even after its discovery, the arguments about the new planet continued to rage. Who, actually, discovered it? That honour went in the end to the men *without* the telescopes – Adams and Le Verrier, who had predicted where the new world would lie. And then there was the heated debate about what to name it. With an international discovery on their hands, astronomers were in a quandary as to where to go. Was it to be Herschel? Janus? Oceanus? Or Le Verrier?

One of Airy's colleagues – Airy, of course, having egg all over his face for abjectly failing to discover the new world in Britain – made his point to the Astronomer Royal: 'Mythology is neutral ground. Herschel is a good name enough. Le Verrier somehow or other suggests a *fabriquant* [manufacturer] and is therefore not so good. But just think how awkward it would be if the next planet should be discovered by a German, by a Bugge, a Funk, or your hirsute friend Boguslawski!'

In the end, reason prevailed. The bluish-green disc of the new planet brought to mind an aqueous world – and the planet was named by Le Verrier after the god of the sea, Neptune.

Now, our Solar System of the Sun and its eight planets was complete – or was it?

Chapter 7

THE PLANET THAT NEVER WAS

Alan Stern is a man with a mission. It's a mission that has taken him fourteen years to fulfil – a mission that humankind has never attempted before. One that has sent a space probe to the very edge of the Solar System. On 14 July 2015, the spacecraft *New Horizons* will speed past its target – a controversial world commanding our Sun's very frontiers.

Even after the discovery of Neptune, astronomers were uneasy as to whether they had located the outermost planet in the Solar System. Soon after Galle's pinpointing of the new world, Le Verrier was still worried about persistent discrepancies in the orbit of Uranus. Was there another world beyond Neptune?

The scene shifts to America, where a supremely talented banker and polymath strides onto the scene. Percival Lowell was born into a wealthy Boston family on 13 March 1855. The Lowells were second only to the Cabots in the silver-spoon stakes, and it was said:

> And this is good old Boston
> The home of the bean and the cod

Where the Lowells talk only to Cabots
And the Cabots talk only to God.

These lines of doggerel were penned at Harvard University, where young Lowell graduated with a distinction in mathematics. His acceptance speech was deemed to be far beyond his years – an analysis of the 'nebular hypothesis': the theory as to how the planets were created.

Lowell went on to pursue a glittering international business career, but would forever be drawn back to his first love – astronomy. With his considerable wealth, he built a spectacular observatory in Flagstaff, Arizona. Constructed in 1894, it was – at over 2,100 metres above sea level – the world's first high-altitude observatory, where clear skies were the norm.

Lowell had two astronomical passions: Mars, and the search for 'Planet X' – the world he was convinced lay beyond Neptune. In 1906, he began a systematic visual search for 'X', but the field of his telescope was too small. By 1915, he had resorted to maths, but after 100 pages of predictions, he concluded: 'Analytics promise the precision of a rifle, and finds it must rely upon the promiscuity of a shotgun after all ...' Finally, Lowell turned to photography, which by then was a common tool in astronomy. But although he twice managed to image 'X' in 1915, he didn't recognise it – the object was hardly bright enough to be seen.

Lowell died in 1916 – the result of a stroke – and the future of the Lowell Observatory was flung into disarray. Although Lowell provided well for his widow Constance in his will, most of his money was set aside to preserve the future of his observatory. Constance fought the legacy for nearly ten years, until the observatory came into the safe hands of Roger Putnam, a nephew of Percival Lowell and a keen amateur astronomer. Although the Lowell Observatory had lost half of its $2.3m trust funding during the bitter lawsuit, Putnam was determined to revive his uncle's dream. Using a great deal of his own money, Putnam commissioned a new thirteen-inch (33cm) photographic refractor telescope to recommence the search for 'X'. He was keen to follow the observatory's legacy – to discover 'Uncle Percy's Planet'.

The observatory's new director, Vesto M. Slipher, decided to revisit the planet problem in 1929. He realised that he needed someone young and keen who could photograph the sky night after night to check if anything had moved against the distant backdrop of stars. He found his man in 23-year-old Clyde Tombaugh, an amateur astronomer from Kansas, who had just sent Slipher some of his planetary observations.

Tombaugh's excitement was palpable: 'Sending my drawings of Mars and Jupiter to Dr Slipher, Director of the Lowell Observatory, could not have been better timed,' he told us in the 1980s. (Tombaugh sadly died in 1997.)

'Slipher was looking for a suitable amateur to work with the newly acquired thirteen-inch telescope,' Tombaugh recalled. 'He responded almost immediately ... he said that the night work would involve long exposures in an unheated dome. "Would you be interested in coming to Flagstaff on a few months' trial basis, about the middle of January?" Slipher asked.

'My father's parting words were: "Clyde, make yourself useful, and beware of easy women."'

Keen young Tombaugh soon mastered the intricacies of the new telescope, and the blink comparator he would use during the day to compare the photographs. The comparator showed two photographs of the same region of sky in rapid succession. Tombaugh had to set up the photographic plates so that the stars were in the same positions. Then, anything that had moved in the two or three days between the two exposures would appear to jump backwards and forwards, drawing attention to itself.

Slipher suggested that the young man should start his search in the constellation of Gemini, where Lowell had predicted the planet should lie. Lowell had expected a relatively bright new world – one some seven times heavier than the Earth. But Tombaugh drew a blank.

After a few months, he began to take the search into his own hands. Supposing that the new world *wasn't* as heavy as Lowell had predicted; supposing it was a different kind of beast altogether? Tombaugh resolved to search the whole of the Zodiac (the band of the sky that the planets appear to follow), and look for any world beyond the orbit of Neptune.

On 18 February 1930, he compared images taken on the 23rd and 29th of January. Something jumped – just the right amount to be a planet beyond Neptune.

'I was terribly excited,' remembered Tombaugh. 'I don't think I could ever top that thrill.'

Always a perfectionist, Tombaugh checked on other images of the faint object. When he was confident enough, he knocked on the director's door. Taking a deep breath, Tombaugh announced: 'Dr Slipher, I've found your Planet X.' He recalled: 'Slipher rose up with a tremendous look of elation ... I found it hard to keep up with him as he went back to the blink comparator.'

In England, news of the discovery broke in *The Times*. Over breakfast, eleven-year-old Venetia Burney – who was already versed in astronomy and mythology – suggested that the dim world should be named 'Pluto', after the god of the underworld. Her grandfather passed on her suggestion to Oxford's Professor of Astronomy, who sent a telegram to Slipher:

> Naming new planet. Please consider Pluto, suggested by small girl Venetia Burney for dark and gloomy planet.

The name stuck. It also had the added advantage that its first two letters are the initials of Percival Lowell – and a joined 'PL' forms a neat shorthand symbol for the planet.

But Pluto refused to conform. Looking back through old photographs, astronomers found missing images of the new world – allowing them to compute the object's orbit fairly precisely. They discovered that Pluto has the most oval path of any planet, bringing it closer to the Sun than Neptune for part of its orbit, but taking it 60 per cent further out than Neptune when it is at its farthest from the Sun. Although the paths of the two planets cross, there's no chance of a collision; Pluto's orbit is so highly tilted that it passes well above Neptune's path.

And was Pluto the predicted planet that had Uranus in its gravitational thrall? It was so much fainter than Lowell had estimated; it had to be a much smaller world. Even Tombaugh had his doubts. Soon after he discovered Pluto he was back at the telescope, photographing by night and blinking by day. In thirteen years of searching, no new world emerged. But Tombaugh's extended search left a remarkable cosmic legacy. He scrutinised 45 million stars, discovered 775 asteroids, and identified 1,700 stars that vary in brightness.

The only way of pinning down the mass of Pluto was to find a moon in orbit about it. Tombaugh and his colleagues at the Lowell Observatory peered at Pluto with their largest telescopes – and discovered nothing.

All that was to change in 1978, when Jim Christy – a researcher at the US Naval Observatory – was measuring precise positions of objects in the sky on photographic plates captured with a new, very accurate telescope. He was puzzled as to why the images of Pluto appeared pear-shaped, yet those of the stars were perfectly circular. He dumped the plates in a box labelled 'defective'.

But the images still irked him. And there was another possibility. The pear-shape could be the combination of two circular images, one brighter than the other; the brighter image would be Pluto itself, the fainter a moon very close to the planet.

'I didn't really believe it,' Christy admitted. 'I went away and got on with something else, and then later thought: I know it's a moon – I'd better do something about it!'

He checked the other photographs and, sure enough, a high magnification showed that Pluto did appear elongated on other occasions. The elongation of the pear-shape moved gradually around, just as a moon would move round a planet.

As the moon's discoverer, Christy had the privilege of naming it. He wanted to name it after his wife, Charlene, but – as his colleagues pointed out – this hardly followed classical tradition. So one night found Christy poring over a Greek dictionary. To his delight, he discovered that 'Charon' was the name of Pluto's faithful oarsman, who ferried souls across the River Styx to the underworld. 'Char' (pronounced 'Shar') was Christy's nickname for Charlene – so what better name than Charon? In homage to Mrs Christy, astronomers pronounce the moon 'Sharon' (instead of the classical 'Kaaron'). The delighted Charlene told us: 'Many husbands promise their wives the moon, but my husband got it for me.'

In just twenty-four hours, Christy was able to solve the 48-year-old riddle of the mass and gravitational pull of Pluto. Only 2,300 kilometres across, Pluto weighs in at just one five-hundredth the mass of the Earth, and certainly could not pull a heavyweight gas giant like Uranus off course. (Later, in 1989 – when the space probe *Voyager 2* sped past Neptune – astronomers realised that they had hitherto got the mass of this planet slightly wrong. The updated gravitational pull of Neptune entirely explained the discrepancies in Uranus' orbit, which had kicked off the search for Planet X in the first place.)

So Pluto was no more than a tiny cosmic runt – a minuscule world following a wayward orbit, in the twilight zone of our Solar System.

And worse was to come. It took a new generation of astronomers, equipped with better technology, to discover that Pluto was not unique. It had company.

Dave Jewitt – one of the new researchers – remembers growing up in London. 'The only things I could see with my tiny telescope were the Sun, Moon and planets. They basically possessed me, and I've never really shaken them off.' Jewitt was also possessed by 'the apparent emptiness of the outer Solar System'.

Moving to the University of Hawaii in 1988, he and his colleague Jane Luu used one of the largest telescopes on the 4,200-metre peak of Mauna Kea to diligently survey the depths of space for slow-moving objects. Like Clyde Tombaugh, they used a blinking technique to search – but electronics now speeded up the process beyond recognition.

In 1992, the duo hit gold. 'After years of searching, it was a surprise to find something,' Jewitt recalled. What they discovered was the first of the many small, icy bodies beyond the orbit of Neptune that have since been detected.

The idea of a swarm of icy debris at the Solar System's frontier was not new; many astronomers had come up with the suggestion. But the name Kuiper Belt – named after the planetary astronomer Gerard Kuiper – has stuck. More than a thousand KBOs (Kuiper Belt Objects) have since been discovered.

Now, the seeds of doubt were being sown as to Pluto's status as a planet. 'Pluto is an absolutely typical KBO,' remarked Jewitt in 1999. 'I expect that within the next two or three years, we'll find objects that are as big as Pluto, possibly even bigger.'

Enter Mike Brown and his team. An astronomer from Caltech, Brown also sweeps the outer Solar System for tiny, icy bodies, and has a reputation for finding objects that are bigger than other astronomers have discovered.

But in 2003, even he was in for a shock. Searching the frontiers of space with the Samuel Oschin Telescope on Palomar Mountain in California – an instrument that he himself had painstakingly restored – Brown noticed that his computer screen started to come up with images of a very slow-moving object. For two years, the team checked their observations. The object had to be *very* far away. And yet, it was surprisingly bright. Brown's planet-seeker's instincts kicked in: it had to be bigger than Pluto.

Over to the Hubble Space Telescope. It confirmed that the icy body was *probably* slightly larger than Pluto – but only just. Even now, the sizes of the two objects are still under dispute, but Eris, the newcomer, is definitely the more massive. In 2005, the world heard that the solar system had a new planet. Three times further away than Pluto, Brown's object is the most distant known body in the Solar System. For months, the debate raged as to whether there was indeed a Planet 10.

What to call it? Brown – renowned for giving his worldlets crazy names – came up with Xena, the heroine of TV's *Xena: Warrior Princess.* Alas, the powers-that-be weren't impressed. Preferring a conventional choice, they came up with Eris – the Greek goddess of discord.

And what a prophetic name! *Was* Eris Planet 10? Or – as Dave Jewitt had predicted – just another KBO, like Pluto? The argument certainly provoked discord in the world of astronomy. And it made scientists revisit the definition of what, actually, constitutes a planet. In an unprecedented step, the question was put to the vote at the high-level General Assembly of the International Astronomical Union at Prague in 2006.

The result? Pluto was demoted to the status of a 'dwarf planet' – and so was Eris. Mike Brown told reporters, 'Pluto is dead. It is not a planet. There are finally, officially, eight planets in the Solar System.'

So how did Clyde Tombaugh's family take the news? His widow Patricia was contemplative. 'It's disappointing in a way, and confusing ... But I understand that science is not something that just sits there.'

Alden, Tombaugh's son, is very positive about his father's contributions to our knowledge of the Solar System. 'This doesn't change my father's achievement. Science is an evolving process ... and he was part of that process.'

But there is a vocal minority of astronomers who won't take the news of Pluto's demotion sitting down. Chief among them is Alan Stern, a senior planetary researcher at America's Southwest Research Institute in Boulder, Colorado, who has been passionate about Pluto all his life.

His take on the vote? 'Less than five per cent of the world's astronomers voted. It won't stand. It's a farce.'

Stern is principal investigator on the *New Horizons* space probe mission. Launched in January 2006, the craft is scheduled to fly past Pluto and its moon Charon in July 2015. In August 2014, *New Horizons* crossed the orbit of Neptune – coincidentally, twenty-five years to the day after the probe *Voyager 2* flew past the gas giant in 1989.

'*Voyagers 1* and *2* explored the entire middle zone of the Solar System, where the giant planets orbit,' observed Stern. 'Now we stand on *Voyager*'s broad shoulders to explore the even more distant and mysterious Pluto system.'

To Stern and his colleagues, Pluto is far more than a mere afterthought of the Solar System. Much more fundamentally, it represents a new frontier. Pluto, Eris and their companions are part of a huge, unexplored region of our planetary neighbourhood. Stern's ambitious mission is – understandably – exciting him massively. *New Horizons* will swing within 10,000 kilometres of Pluto, and 27,000 kilometres from Charon. The probe is equipped with a suite of experiments designed to image both Pluto and Charon, and to sample their environments.

Does Stern have any expectations? He's open-minded: there is a possibility of discovering rings; perhaps ice-volcanoes like those found by astronomers on Neptune's moon Triton; and maybe unexpected interactions between Pluto and Charon.

Most of all, Stern is anticipating the discovery of a new zone of our Solar System – a region far larger than that we have hitherto explored. 'I don't make predictions. It's all about exploration,' he explains.

And *New Horizons* has plenty to explore beyond Pluto. It has been loaded with enough fuel to make close approaches to other frontier worlds, and those targets are currently under selection. The lure of more and more ever-distant worlds is drawing the imagination of astronomers and space planners ever outwards.

'In a couple of hundred years,' says Alan Stern, 'the Solar System will be open enough and travel will be easy enough for scientists like myself to have the chance to visit these other worlds we are studying. There's a lot of real estate out there: even with very fast transport there's simply so much land and so much variety that it will take a long time to explore it all, to capitalise on it both scientifically and, ultimately, economically.'

But the last word has to go to someone who has already taken the first step on humankind's long march across space.

'We must go there for ourselves,' maintains Gene Cernan, the last man to walk on the Moon. 'The questions about what it is like, what does it feel like, are questions people can only relate to through a human being – not through a robot, not through a camera, not through a computer chip. I mean, no robot has ever had a ticker-tape parade in New York City!

'We are now a truly space-faring people,' he continues. 'Taking a spaceship to Pluto seems almost unimaginable – but, you know, going to the Moon was at one time only a dream, too.'

Chapter 8

SOLAR INFERNO

'Nuclear Power? No Thanks!' Since the 1970s, we've been familiar with this logo, its accompanying happy sunny face, and the message that nuclear energy is bad, while solar energy is natural and good.

But the truth is otherwise. The apparently benign and life-giving Sun is in reality an immense nuclear reactor, sitting right on our cosmic doorstep. Every second, the Sun is processing more nuclear fuel than every power station on Earth will consume in millions of years. And it is emitting radiation so lethal that it would kill everyone on our planet, if we weren't protected by the Earth's magnetic umbrella and the invisible shield of our atmosphere.

The Sun is the most familiar object in the whole sky, so it comes as something of a shock to realise just how little astronomers knew about it until very recently. We still can't predict the Sun's erratic outbursts, which could – without warning – destroy our civilisation in seconds.

Scientists didn't give much thought to this immense inferno in the sky until the middle of the nineteenth century. And the person who prompted them into action wasn't an astronomer, or even a physicist – but the biologist Charles Darwin.

In 1859, Darwin had tossed a fire-cracker into the staid Victorian world, with the outrageous claim in his book *On the Origin of Species* that all life on Earth had evolved from simpler ancestors. The contentious book famously got him into hot water with the Church. But Darwin's idea also riled some of the leading physicists of the day.

Darwin realised it must have taken a very long time for species of life to evolve, so the Earth must be much older than most people imagined. To work out the age of our planet, Darwin investigated the erosion of the rocks forming the Weald district of Kent. From the speed at which cliffs were being worn away in his day, Darwin concluded: 'the denudation of the Weald must have required 306,662,400 years; or say three hundred million years'.

That was impossibly long for many scientists to believe. Darwin's principal critic was the physicist Lord Kelvin. The Earth couldn't be any older than the Sun, Kelvin rightly argued, and so he set about working out the Sun's age. Scientists knew how hot the Sun was, and how rapidly it was beaming away its energy. Clearly something was fuelling its fires: if Kelvin knew what, he could calculate the Sun's lifetime.

The American astronomer – and aviation pioneer – Samuel Pierpoint Langley had already checked out whether the Sun could be a great lump of burning coal. But his calculations fell woefully short: 'it would require all the coal of all the coal fields of Pennsylvania to keep up the energy of the Sun for one-thousandth of a second'. Even if the whole Sun were made of coal, it would burn out in only a few thousand years.

Inspired by the spectacular sight of incandescent shooting stars streaking through the Earth's atmosphere,

Lord Kelvin at first believed that infalling meteors kept the Sun hot. But he deduced it would mean a mass of meteors as great as the Earth impacting the Sun every forty-seven years. And that would have seriously upset the orbits of the planets.

So Kelvin settled on an idea first proposed by the German scientist Hermann von Helmholtz. A gas becomes hotter when you compress it, and Helmholtz suggested that the Sun's gravity is squeezing it to produce heat. Kelvin did the sums on this 'contraction hypothesis', and announced that the Sun had existed for 20 million years – not nearly long enough for life to evolve. Trumpeting the victory of physics over geology, he asked rhetorically: 'What then are we to think of such geological estimates as 300,000,000 years for the "denudation of the Weald"?'

As Kelvin intended, Charles Darwin was intimidated by this blast of cold science. He deleted his geological calculations from later editions of *On the Origin of Species*, saying that Kelvin's 'views on the recent age of the world have been for some time one of my sorest troubles'.

In hindsight, we know that when it came to the age of the Sun – and of the Earth – the physicist was wrong, and the biologist was right. In fact, both are even older than Darwin had suspected. The Sun has actually been shining for at least ten times longer, some 4,567 million years.

The mystery of the Sun's heat was eventually solved by two of the greatest scientific minds at the start of the twentieth century. In 1905, Albert Einstein in Switzerland deduced that matter can change into energy – expressed pithily in the famous equation

$E=mc^2$, where 'E' is the energy that can be obtained from a mass 'm', and 'c' is the speed of light.

Fifteen years later, this iconic equation led British scientist Arthur Eddington to another 'eureka' moment. First, he roundly derided Kelvin's theory: 'Only the inertia of tradition keeps the contraction hypothesis alive – or rather, not alive, but an unburied corpse.'

Eddington instead focused on the hydrogen gas in the Sun. In theory, you can slam together four hydrogen atoms to make the next atom, helium. But a helium atom is slightly lighter in weight than four hydrogens: what happens to the mass that goes missing?

Einstein's famous equation held the answer: the missing mass turns into the energy that makes the Sun shine. According to Eddington, the source of the Sun's power 'can scarcely be other than the subatomic energy which, it is known, exists abundantly in all matter; we sometimes dream that man will one day learn how to release it and use it for his service.'

In a lecture that must count as one of the most amazing talks in the history of science, Eddington went on to say that there was enough hydrogen in the Sun to maintain its output of heat for 15 billion years – plenty of time for life to evolve on Earth. And – presciently – he concluded, 'it seems to bring a little nearer to fulfilment our dream of controlling this latent power for the well-being of the human race – or for its suicide'. Only a year after the formal end of the First World War, Eddington was predicting the earth-shattering power of the hydrogen bomb!

And he was right: the Sun is, indeed, an unimaginably colossal H-bomb that – fortunately for us – is running in slow-motion. The vast weight of the Sun's outer regions prevents the core from erupting in a devastating explosion that would certainly contradict the 'Nuclear Power? No Thanks!' message.

At the Sun's core, conditions are extreme beyond our imagination. Gravity squeezes so tightly that the gas here is seventy times denser than gold. The temperature soars to 15.7 million degrees Celsius. Nuclear reactions are changing hydrogen into helium at such an incredible rate that the Sun is destroying four million tonnes of its matter *every second* – and converting it into pure energy.

Yet here's a sobering thought. If you cut out a region of the Sun's core as large as yourself, it would actually be producing less energy than your body generates. The Sun has such a vast power output only because its central reactor is huge – big enough to contain thousands of Earths.

This energy floods out of the core as lethal radiation. If we were directly exposed to these deadly gamma rays, we wouldn't survive for a second. Luckily for us, the Sun's nuclear reactor is safely cloaked by almost a million kilometres of dense gas. As the radiation crashes into this thick fog, it's constantly diverted into new directions, in what scientists call a 'random walk'. It's like a drunk trying to make his way from a lamp-post up the street; he takes each step in a random direction, so – overall – he hardly makes any progress.

As a result, the radiation from the Sun's core bounces around inside our local star for about 100,000 years. When it finally emerges from the solar surface, the energy of the gamma rays has been tamed: the Sun's awesome power pours out as heat and light. And the slow random walk means that the light we see from the Sun today was created in its core before our early human ancestors had left Africa! After that, it takes a mere eight minutes for the Sun's light and heat to speed through space to the grateful recipients on planet Earth.

Without our star's warming rays, life on Earth would be impossible. So it's appropriate that humans have worshipped the Sun from the very earliest times. Archaeologists digging at Jinsha, in China, have unearthed an ancient gold disc that shows the Sun surrounded by four birds which – according to legend – pulled it down to the horizon at sunset.

According to the *Astrological Treatise* of the Chin Dynasty: 'The Sun, as essence of Mature Yang, governing all life, sustenance, benevolence and virtue, is the counterpart of the Ruler of Men.' It goes on to say that if the Ruler of Men – the Chinese Emperor – has any flaws, the Sun will reveal them.

For that reason, the Chinese kept the Sun under close scrutiny. And, indeed, they did – from time to time – notice telltale blemishes, which we now call sunspots. 'When a crow appears within the Sun, the ruler is not enlightened, and his governance is chaotic.' And that was just the beginning: 'When in the Sun there are black spots, now a couple, now several, vassals will set aside their lord.'

Recently, astronomers have discovered that these dark sunspots do indeed hold malevolent and disruptive powers. The first to experience them at first hand were Britain's radar operators, in the depths of the Second World War.

It was 1942, and the Allies were fearful of a major new German weapon: radio transmitters on their bombers that would 'jam' the British radar. Sure enough, on 27 and 28 February, radar sets along the English Channel were overwhelmed by powerful signals that prevented them from detecting anything.

'There was much alarm wondering what this unknown radiation might portend,' recalled Stanley Hey in his autobiography, *The Secret Man.* A young physicist – seconded from his position as a teacher at Burnley Grammar School – Hey was working with the British Army on radar jamming.

As the military top-brass panicked, Hey studied the reports from the radar stations. He found that the 'jamming' began at sunrise, and ended at sunset. In addition, the general direction of the radio signals followed the motion of the Sun across the sky, from east to west.

'I telephoned the Royal Observatory,' Hey continued, 'and enquired whether there was anything unusual about the Sun. The answer came that a very large sunspot was present near the centre of the solar disc.'

The Germans' 'secret weapon' was in fact a stupendous blast of radio waves unleashed from the Sun's surface, 150 million kilometres away. The source was the magnetic maelstrom of a sunspot.

Astronomers in the West stumbled over these dark markings when the telescope was invented. They had to project the Sun's image onto a screen, because the telescope concentrated its heat so much that it would blind anyone who looked directly at our local star. (This warning must still be heeded today: if you want to observe sunspots yourself, you must project the Sun's image, or use a metallised-plastic filter from a reputable supplier.)

According to legend, the Italian pioneering scientist Galileo was the first to observe sunspots. In fact, the

British astronomer and explorer Thomas Harriot drew dark markings on the Sun a year earlier, on 8 December 1609. But Galileo did prove to his own satisfaction that a sunspot is a blemish on the Sun itself, rather than the silhouette of a small planet passing across the Sun's face.

Two centuries later, the hope of discovering the dark shape of a new planet led German apothecary – and amateur astronomer – Heinrich Schwabe to study the Sun on every clear day. But he quickly discovered that his path to fame was strewn with misleading sunspots. To eliminate these false alarms, he began to plot their positions.

Planet-hunting quickly faded from Schwabe's mind, as he became obsessed with sunspots themselves. When Schwabe was ill, he asked his family to bring his telescope to his bedroom; when he travelled, he persuaded them to sketch the Sun for him. Over the course of 43 years, the indefatigable sun-worshipper logged an incredible 134,000 measurements of sunspots.

And he noticed something rather unusual. 'From my earlier observations,' Schwabe reported, 'it appears that there is a certain periodicity in the appearance of sunspots and this theory seems more and more probable from the results of this year.'

Schwabe had discovered the 'sunspot cycle' – a regular ebb and flow in the number of spots disfiguring the Sun's face. They come and go over a period of roughly eleven years.

The driving force behind the sunspot cycle is the Sun's magnetism. Within our star's rotating gaseous bulk, magnetic fields wind up – like a rubber band being stretched around a spinning ball. After eleven years, bundles of magnetic field break through the glowing surface, as dark sunspots. Over the next few years, these loops of magnetism settle down, and the Sun becomes quiet again.

Sunspots are awesome. The largest are bigger than our planet Earth. They're cooler than the rest of the Sun's surface – but 'cool' still means a temperature of 4000°C. And though sunspots look black, that's only because they are dimmer than the brilliant solar surface: if you could see a sunspot in isolation, it would shine as brightly as the full Moon!

The Sun's magnetism rears upwards over sunspots, in vast loops that are outlined by faint glowing gas, like the iron filings we use in school to show the magnetic field around a magnet. There's only one time you can see these gorgeous magnetic details in the Sun's tenuous atmosphere with your own eyes: when the blindingly bright surface is blocked from view, during a total eclipse of the Sun.

We've been lucky enough to witness half a dozen total eclipses; and we can guarantee it's the most weird sky-sight you'll ever see – definitely one for your 'bucket list of things to see before you die'.

For our first eclipse, we travelled to a beach on the remote Indonesian tin-mining Bangka Island. A bunch of astronomers from Europe and America set up cameras, telescopes and electronic equipment facing out to sea, where the Sun was rising – oblivious to the fate in store for its glowing orb. Behind us, the locals were gathering in anticipation too: not only to view the eclipse, but to see for themselves the pale 'Moon-faced' people they'd seen on television, but not in the flesh! An eclipse is a unique forum for bringing together – in shared wonder – humans from the most varied cultures.

Peering through our solar filters, we viewed the dark silhouette of the Moon encroach on to the solar disc. As the Sun's smiling face shrank to a thin crescent, the lighting around us changed. The landscape appeared greyer and flatter as the Sun's light shrank to virtually nothing. The warm tropical air around us chilled in anticipation.

Up in the sky, we glimpsed the first faint wisps of the Sun's atmosphere, even before its surface was totally hidden. Then, suddenly …

In the sky, there's a completely unfamiliar – and frightening – celestial object. Instead of the brilliant Sun, we were confronted by a powerful new celestial being. In Heather's notes from the time, it's 'a powerful king wearing a war-mask – almost like a deity demonstrating its power'.

Nigel saw an oriental dragon, its black mouth surrounded by golden fronds. He wondered if that's what gave rise to legends in the Far East that the Sun was eclipsed when it was swallowed by a dragon. The black dragon's mouth was edged with gaudy red lips – the lower part of the Sun's atmosphere, known as the chromosphere for its intense colour.

Time to swing into action with the binoculars: a total eclipse is the *only* time it's safe to turn binoculars or a telescope to the Sun. Heather recorded: 'the edge of the disc is fringed with the deepest crimson prominences imaginable – lots and lots of them! One of them is enormous: a great double loop.'

With the naked eye, we could easily see the outer regions of the atmosphere, the faintly glowing corona: 'around the chromosphere, the corona extends like a lion's mane into a deep purple sky'. Fluted lines of luminous gas stretched outwards for well over the diameter of the Sun itself.

The cerise prominences in the chromosphere, and the glowing tendrils in the corona above, were the visible signs of the Sun's magnetism reaching outwards, and sculpting its thin but ultra-hot atmosphere. But, for the moment, scientific excitement had to yield to sheer awe.

The instant the Sun disappeared, our ears were assailed by cries and yells. Even seasoned eclipse veterans were saying, 'I can't believe it.' The sheer power and majesty of the eclipsed Sun is quite indescribable.

After a couple of minutes – which sped by all too quickly – another miracle appeared: a celestial diamond ring. The first speck of the Sun's brilliant surface reappeared as a dazzling gem set in the gently shining band that surrounded the Moon's dark shape. The Sun's light flooded back – to reveal a crowd of astounded faces lining the tropical shore.

Euphoria reigned. Seasoned astronomers and local Indonesians hugged spontaneously with joy. We agreed that eclipse-fever is a serious affliction – and we began to plan when and where we'd need to travel for our next bout!

Some professional astronomers still voyage long distances to view eclipses when they need to study details of the Sun's magnetic atmosphere. But it's not obligatory any more. Researchers have launched space telescopes tuned to natural X-rays – such as NASA's Solar Dynamics Observatory – which can reveal its pulsating and undulating loops in exquisite detail.

This seething riot of activity reaches a climax when two magnetic loops touch. They short-circuit in a solar flare – a searingly hot explosion that beams deadly radiation into space at the speed of light.

The turbulent magnetic inferno in the Sun's atmosphere can also unleash huge clouds of hot gas into space. Inelegantly known as 'coronal mass ejections', these solar superstorms take longer to cross interplanetary space and hit the Earth: but they can pack an even bigger wallop than a solar flare.

Fortunately, our planet has a sturdy umbrella. The Earth's magnetism deflects the gas clouds towards the poles, where they stream downwards to impact our second line of defence – the blanket of air surrounding the planet. The energy of the solar storms lights up the upper regions of the atmosphere, as the phantasmagorical Northern and Southern Lights.

The aurora is a great sky-sight – like a solar eclipse – and is well worth a special expedition. The best viewing places are near the Earth's poles: in Finland, you can even rent a transparent igloo where you lie in bed and view the Northern Lights unfolding overhead!

The Northern and Southern lights appear as dazzling curtains and streamers of green light, fringed with shades of red and purple, rippling and dancing their way across the sky. If you're right underneath, the aurora fans outwards from the zenith, turning the entire sky into a pulsating flower that's blossoming above you.

When the Sun is deeply troubled, the aurorae spread from their usual haunts towards the Equator. In September 1859, the *New York Times* reported: 'strange fires overran the entire heavens – now separating into streamers, gathered at the zenith, and forming a glorious canopy – then spreading evenly like a vapor, shedding on all things a soft radiance; again, across the sky waves of light would flit, like the almost undistinguishable ripple produced by the faintest breeze upon the quiet surface of an inland lake.'

This incredible phenomenon was visible as far south as the Caribbean. Astonished inhabitants of Jamaica saw the reddish glow of the aurora to the northeast, and thought that Cuba was on fire.

But the beautiful skies were matched by utter mayhem in the offices of the telegraph companies – the internet of their day, carrying long-distance messages tapped out in Morse code. That night, telegraph operators found that vast electric currents were sweeping through the wires. They tried to protect their batteries by unplugging them – but the natural electricity surge provided enough power to transmit messages across the width of the United States.

Where chemically coated paper was used to record the incoming stream of dots and dashes, sparks flew out of the equipment and ignited the charts. In Washington, DC, telegraph operator Frederick Royce reported: 'I received a very severe electric shock, which stunned me for an instant. An old man who was sitting facing me, and but a few feet distant, said that he saw a spark of fire jump from my forehead.'

Just before all this mayhem erupted, an English amateur astronomer and brewer, Richard Carrington, had spotted a solar flare so extraordinarily bright that it was visible in an ordinary telescope. Astronomers now calculate that this magnetic eruption on the Sun exploded with a force 10 billion times greater than the nuclear bomb that destroyed Hiroshima in 1945.

The colossal flare – now known as the Carrington Event – unleashed a superstorm of electrically charged particles that smashed into the Earth's magnetic field. Our planet's invisible umbrella swept most of the dangerous particles towards the poles, lighting up the aurorae. But the force of the impact stretched and buckled the invisible lines of magnetism. The writhing magnetic field generated huge currents in the Earth below – especially in the long electrically conducting wires that stretched out between the telegraph stations.

Before the Victorian era, people would only notice such a solar eruption by the brilliant lights in the sky. In 1859, the fledgling communications system was hard hit. And – as the world has come to depend ever more on electricity and electronic communications – there's a growing danger that a solar superstorm will bring our civilisation to its knees.

'Today's world would be severely damaged if a Carrington Event were to repeat itself,' says Ashley Dove-Jay, a British engineer who has assessed the dangers with an international group of space experts. 'The consequences could be catastrophic and long-lasting.'

The first effect you might notice is that your mobile phone stops working. The superstorm won't damage the handset itself; but it will disrupt the signals passing from one cell-tower to the next. Power surges could even destroy the communications towers.

And that's just the beginning. Massive currents would overload the electricity grids that now stretch across continents. We had a warning in March 1989, when a smaller superstorm caused a major blackout in Canada, plunging six million people into darkness, cold and confusion for nine hours.

A full-scale Carrington Event could destroy the high-voltage transformers that link sections of the electricity grid. These could take a year to replace. When the grid fails, cooling systems will collapse at nuclear power stations – which may then suffer catastrophic meltdown. Without electricity, the internet will go down. Pumps at petrol stations will stop working, so deliveries will grow increasingly difficult, and supplies will run out at shops. Water and sewage systems will struggle; and epidemics of disease may sweep across cities.

The impact will be felt least in India and China, because they are farthest from the Earth's magnetic polar regions and have simpler and more robust electricity grids. These countries also have soils that are less good at conducting electricity, again minimising the damage. Dove-Jay predicts that India and China may have to help the rest of the world to recover.

'Space weather destroys stuff,' concurs NASA space expert Pete Worden, who launched an observatory in 2013 to probe the Sun's magnetism. 'It's important we learn how to live with our star.'

Worden takes one particular example to show that these doom-laded predictions are not just scaremongering. On 23 July 2012, satellites saw the Sun erupt with a Carrington-strength explosion: fortunately for us, the superstorm blasted out through the Solar System in a different direction. Had it occurred a week earlier, our planet would have been right in the firing line.

And, over enough time, the Sun will fire outbursts even more powerful than the 1859 shocker. Scientists studying tree-rings from Japan have found that ancient cedar trees were blasted by radiation from space in AD 774 – most likely from a solar explosion ten to twenty times stronger than the Carrington Event.

The historical records are sparse and patchy, but astronomers can work out roughly how often we'll be subjected to the Sun's full fury by checking out giant flares on other stars – the Sun's siblings. They are digging into data from the *Kepler* space telescope, whose main mission is to hunt for planets orbiting other stars. The answer is sobering. Every 500 years, our planet is likely to be hit by a solar superstorm 100 times stronger than the Carrington Event – with rarer eruptions that are ten times more powerful still. These could destroy civilisation as we know it.

What can we do to protect ourselves?

'We cannot control the mood of our nearest star,' Dove-Jay muses. His team suggests sending a swarm of small spacecraft to orbit close to the Sun's surface, checking out its magnetic weather at first hand. This system could provide a week's warning of a solar superstorm, allowing us to switch off vulnerable electricity grids and put into action emergency plans for fuel, food and medicine.

Thanks to several asteroids narrowly missing the Earth recently, people are now well aware of the dangers of a giant space rock impacting our planet (more details coming up in Chapter 15). And films like *Armageddon* have shown how we could – possibly – save ourselves, by sending an astronaut on a derring-do mission to blow up the errant asteroid.

Yet, for every asteroid strike that causes – say – $10 trillion damage, our planet will be buffeted by ten solar storms that create the same impact on the world's economy. As far as cosmic threats go, the Sun is the elephant in the room that people and governments are not taking seriously enough.

'Hollywood has incredible power,' observes Dove-Jay. 'To raise public awareness and therefore government funding dedicated to our preparedness, I think we need a Bruce Willis film on solar superstorms!'

Chapter 9

CAULDRONS OF THE COSMOS

> To see a World in a Grain of Sand
> And a Heaven in a Wild Flower,
> Hold Infinity in the palm of your hand
> And Eternity in an hour.

So wrote William Blake in his 'Auguries of Innocence' … but these words could easily refer to the way in which we see the lives and deaths of the stars today.

While we live our allotted threescore years and ten, stars exist for billions, or even trillions, of years. However, it's amazing to reflect that – in the mere snapshot of time that is the duration of a human life – we have been able to piece together the lengthy, convoluted life-stories of the stars.

The key to unravelling the secret of star-life is that we're fortunate to be blessed with millions of these celestial gems, all in different stages of their evolution. And that's how astronomers have got to grips with the birth, life and death of a star.

Imagine: you're a Martian with just five minutes to spend on Earth. You land your UFO in – say – Oxford

Street, or Fifth Avenue. The streets are swarming with a multitude of people, with a huge range of appearances. But it wouldn't take our intelligent Martian long to work out that they look different because they are of different ages. There are babies in buggies; cool, street-smart kids; young professionals; and slower-moving, elderly people. The whole of a human lifespan is paraded in front of you.

So it is with the stars. But only two centuries ago, the 'official' line was that we were destined to know almost nothing about our celestial companions. The French philosopher Auguste Comte proclaimed: 'We can imagine the possibility of determining the shapes of stars, their sizes or movements; whereas there is no means by which we will be able to examine their chemical composition ...'

Unbeknown to Comte, a revolution was under way. Astronomers were already equipped with telescopes powerful enough to measure the distances to the stars, so pinning down their properties. And a young Bavarian glassmaker, Joseph von Fraunhofer, had been making the most exquisitely crafted glass for telescope lenses. He also produced triangular prisms of glass, which split up 'white' light into a spectrum (like the colours of a rainbow). He thought his craftsmanship was flawed when he discovered that the spectrum of the Sun – and the spectra of several stars – were crossed by dark lines.

Fraunhofer was dismayed – and baffled. 'In all my experiments I could, owing to lack of time, pay attention only to those ... which appeared to have a bearing on practical optics,' he lamented.

The scene now shifts to the University of Heidelberg in the 1850s. The chemist Robert Bunsen (yes: he of Bunsen burner fame) and his physicist colleague Gustav Kirchhoff were intrigued that Fraunhofer had also conducted his prism experiments on flames. The duo noted that Fraunhofer logged *bright* lines in the positions of the dark lines he had observed in the Sun and stars. Using the Bunsen burner, they re-created his experiments, heating up different elements – sodium, lithium, potassium and calcium – to incandescence.

Each element displayed a uniquely different pattern of lines. Then the cosmic fingerprinters went a step further. Kirchhoff discovered that the bright emission lines could be converted into Fraunhofer's dark stellar absorption lines by passing a bright background light through the flame.

Bunsen realised the implications immediately. 'At the moment, I am occupied by an investigation with Kirchhoff which does not allow us to sleep. Kirchhoff has made a totally unexpected discovery, inasmuch as he has found out the cause for the dark lines in the solar spectrum and can produce these lines artificially intensified both in the solar spectrum and in the continuous spectrum of a flame, their position being identical with that of Fraunhofer's lines.'

In other words, the dark absorption lines from the Sun and stars came from a cooler atmosphere wrapped around a hotter globe. Bunsen continued: 'Hence the path is opened for the determination of the chemical composition of the Sun and the fixed stars.'

The two cosmic forensic scientists had invented the world of spectroscopy: a world that was to prove Comte completely wrong, and one that would pave the way to studying the make-up of stars millions of light years away. Fraunhofer's offending dark lines were named in his honour – and what a window they have proved to be in opening up the mysteries of the universe.

Spectroscopy dominated astronomy in the late-Victorian era. Wealthy amateur astronomers took delight in building observatories to analyse the composition of the stars. But the fundamental discovery fell to an astronomer from the tiny market town of Wendover, in beech-clad Buckinghamshire.

Cecilia Payne-Gaposchkin was born in 1900: a sighting of a meteor at the age of five would dictate her life. She was determined to become an astronomer, but later confessed she'd been worried that 'everything might be found out before I was old enough to begin'.

Fortunately, there was plenty for Payne-Gaposchkin to find out. Unbelievably, for a woman of her era, she managed to study for a doctorate at Harvard University in the field of spectroscopy. Astronomical historian Owen Gingerich – who attended her courses and knew 'Mrs G' well – recalls her thesis with reverence. '*Stellar Atmospheres* was undoubtedly the most brilliant PhD thesis ever written in astronomy.'

The first student – male or female – to earn a doctorate from the Harvard College Observatory, Cecilia Payne-Gaposchkin discovered one of the most fundamental factors underpinning our universe. From her researches in spectroscopy, she found that hydrogen was way and ahead the most abundant element in the cosmos.

To begin with, astronomers weren't willing to side with the young graduate student's conclusions: after all, they were used to the composition of the Earth – rich in heavy elements, like iron. But soon they were won over by her evidence.

In 1976, just three years before her death, Payne-Gaposchkin won the prestigious Henry Norris Russell Prize from the American Astronomical Society. In her acceptance speech, she told the audience: 'The reward of a young scientist is the emotional thrill of being the first person in the history of the world to see something or to understand something. Nothing can compare with that experience.'

By catapulting hydrogen to a pre-eminent position in the cosmos, Payne-Gaposchkin had prepared the ground that would now allow astronomers to understand how the stars are born, live and die.

Space is far from being a vacuum. Instead of being empty, it's threaded with the raw materials of starbirth: hydrogen gas in abundance, and dark particles of 'cosmic soot' – dust grains that have evaporated from the surfaces of cool stars.

Over aeons of time, gravity curdles the nascent star-stuff into enormous dark clouds. The archetype is the 'Coalsack' in the constellation of the Southern Cross, which in Aboriginal culture forms the head of the 'Emu'. Orion's Horsehead Nebula looks like a small, but perfectly formed chess-piece. And, on the grandest scale, the Great Rift in Cygnus appears to tear the Milky Way apart.

These looming clouds are impenetrable; but astronomers have ways of piercing the murk. The secret is to look at these stellar nurseries in a new light: infrared.

Infrared is an upmarket term for heat radiation. It's what happens when you turn on an old-fashioned electric fire: you feel the warmth from the element before its red glow kicks in. It was another discovery made by

William Herschel (of Uranus fame), who found that if you spread out sunlight into a spectrum and placed a thermometer beyond the red end of the sequence of colours, it registered an increase in temperature. Herschel concluded that he was witnessing the effects of invisible radiation: 'below red', or 'infrared'.

And so it is inside the dark clouds. With no opposition, gravity is taking over. The cloud breaks up into separate dark fragments, each shrinking under its own inexorable pull. These infant 'protostars' warm up as they contract, just as the air in a bicycle pump gets hotter as you compress it to inflate a tyre. The infrared radiation given out by these fledgling cocoons of gas is a sure sign that starbirth is imminent.

The protostars may be securely hidden in the dark nests – but it's far from being a pleasant place to come into the world. Astronomers call these dark dusty regions 'molecular clouds'; and for a good reason. As well as being hugely rich in hydrogen gas, they're packed with molecules that would be the envy of any well-stocked chemistry lab. And what a smelly bunch they turn out to be! Pungent ammonia, bad-eggs hydrogen sulphide, and the mothballs stench of naphthalene all mingle together. Wafting around in the dark recesses of the cloud are poisonous molecules of cyanide. And – lacing this celestial chemical cocktail – are copious quantities of alcohol. All this heady mix is poised to be incorporated into new stars.

And the happy event is now only a moment away. Gravity is still in control of the collapsing fragments; but it has another competitor: the heat being generated inside the shrinking protostars. When the temperature of our protostar's core reaches 10 million Celsius, a new contender kicks in; and it is the point of no return. The protostar's centre is so hot that the gas is forced to undergo fusion reactions. Its nuclear fires ignite, energy floods through, and the inpull of gravity is – temporarily, at least – vanquished.

A star is born.

A baby star is a violent beast. It shoots out vicious beams of hot gas from its poles, destroying the dark cloud that created it. As the shreds of the gas cloud disperse, the young star finally emerges to greet the gaze of the universe.

Newborn stars don't face their debut alone. Because of the way in which its parent cloud broke up during its early collapse, a young star is born with a host of brothers and sisters. They are clustered tightly together, like a clutch of eggs in a nest.

All around them are the rags and tatters of their natal gases – but now they are decked out carnival-fashion. Gone forever is the brooding dark cloud. In its place is a brilliantly glowing nebula, shocked into shining by powerful ultraviolet radiation from its youthful progeny inside.

Nebulae are among the most glorious, and most-photographed, objects in the universe. The Orion Nebula, the Rosette, the Lagoon, the Trifid ... the list is endless. They look like serene crucibles of gentle starbirth. But such beauty conceals the fact that the violence is still continuing. Even as we watch, hurricane-force stellar winds are tearing apart these incandescent glories of the cosmos. In just a few moments of cosmic time, they will be gone, their gases dispersed into space.

Meanwhile, the young stars are growing up. Gravity still has the upper hand on them: this time, by holding the nestlings together for as long as possible. The infant stars bundle together in star clusters. Most famous is the lovely Pleiades: the Seven Sisters. People usually see any number of stars *but* seven: it's usually six – but some sharp-eyed observers can spot eleven.

Alfred, Lord Tennyson, described the Pleiades beautifully in his poem *Locksley Hall*:

> Many a night I saw the Pleiads rising thro' the mellow shade,
> Glitter like a swarm of fire-flies tangled in a silver braid.

But even star clusters don't last forever. Eventually, the stars break the bonds of gravity and pursue their own paths around the Galaxy. Each star is now a fully paid-up cosmic cauldron in its own right.

For most of its life a star is powered – like our Sun – by fusing hydrogen into helium at its core. Each time, a little mass turns into the energy that makes the star shine. Stars are the natural nuclear reactors of the universe.

After the tribulations and turmoil of starbirth, a star settles down into the prime of its life – the 'Main Sequence'. This is by far the longest part of a star's lifetime; and it can be pretty uneventful. But stars are never boring: even in maturity, they are a diverse bunch. Like human beings, stars range widely in size, weight, lifespan – and even in their personalities!

Just look at the stars on a dark transparent night. Let your eyes get used to the dark, and scrutinise their colours. You'll notice that not all of them are white. In the constellation of Orion alone, there's baleful red Betelgeuse; now – contrast it with steely blue-white Rigel in the same constellation.

Star colours are a stellar thermometer. The hottest stars are blue-white: Rigel boasts a temperature of 12,000C. Next come white stars, followed by yellow (like our Sun), orange, and red. The temperature of Betelgeuse struggles to make it to 3,500C. It's amazing to think that – with just your naked eye – you can take the temperature of a distant star hundreds of light years away!

Many stars are not singletons, but double: the pair cling together from birth. In the Plough lies the famous double star Mizar and Alcor – 'the horse and rider' – which is easily visible to the unaided eye. Castor, in Gemini, is a sextuplet of stars in a cosmic embrace.

Other stars change in brightness. They swell and shrink, varying in their output of light and heat – which must be a nightmare situation for any encircling planets.

But the key to a star is how heavy it is. Mass is everything. Heavyweight stars live fast and loose, and will die after only a few million years. Middling stars, like our Sun, can hang around for billions of years. Low mass stars can go on for trillions of years.

Spica – the brightest star in Virgo – is enormous. Ten times heavier than the Sun, it's 12,100 times more brilliant. And its surface seethes at a temperature of 22,400C (as compared to the Sun's 5,500C).

Contrast this monster with our nearest stellar neighbour: Proxima Centauri. This tiny red star – just one-seventh the mass of our Sun – still packs a fierce magnetic punch, generating giant flares. But it's only a thousandth as bright as our local star, and so energy-conservative that it will last another 400 trillion years.

However magnificent – or humble – the star, the end must eventually come. The supply of hydrogen fuel in

bove: The Sun sets dramatically behind Stonehenge on Midwinter's Day. According to the ıtest evidence, the great stone circle was built to celebrate the winter solstice, not the middle of ummer. *© Stephen Dorey/Alamy*

left: These corroded bronze gearwheels are the remains of the world's first computer – the Antikythera Mechanism – a two-thousand-year-old astronomical calculator constructed by the ancient Greeks. *© Thanassis Stavrakis/AP/Press Association Images*

right: The eleventh-century Persian polymath Omar Khayyam is famed for his poetry celebrating wine, women and song. Also a leading astronomer, Khayyam devised a calendar unsurpassed in accuracy even today. *© Lebrecht Authors/Lebrecht Music & Arts/Lebrecht Music & Arts/Corbis*

left: Eccentric aristocrat Tycho Brahe – great measurer of the cosmos – commands his servants to move his instruments at his manor house at Hven, Denmark. Unperturbed, Tycho's dog sleeps peacefully at his master's feet.

© DEA/G. Dagli Orti/De Agostini/ Getty Images

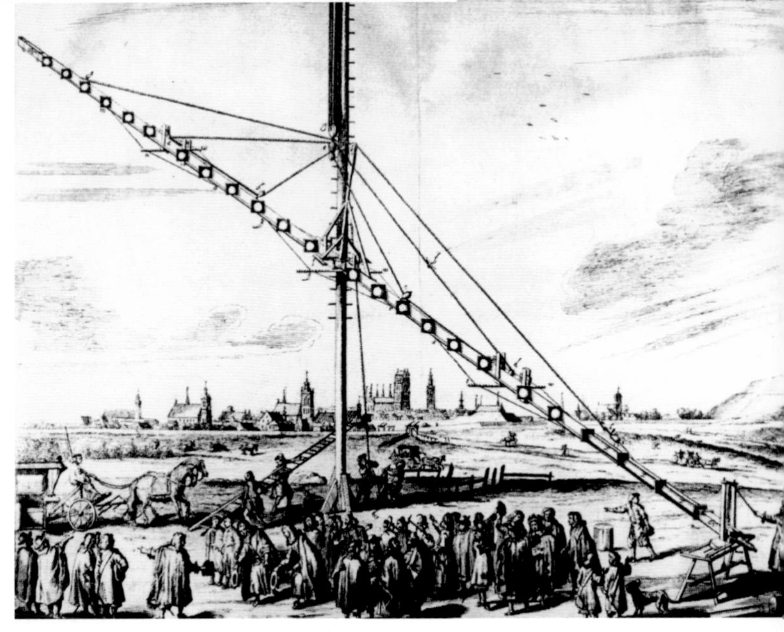

right: Johannes Hevelius built this hugely extended telescope – 46 metres (150 feet) long – at Danzig (now Gdansk in Poland) in the seventeenth century, to minimise distortions caused by its primitive front lens.

© Universal History Archive/Getty Images

above: When the space probe Voyager 2 swept past gas-giant Neptune in 1989, it discovered a blue world wracked by storms. The Great Dark Spot was as big as the Earth, but has since faded. *© 1999 Topham Picturepoint*

right: A young Clyde Tombaugh poses proudly at the business-end of the telescope with which he discovered Pluto, at the Lowell Observatory in Flagstaff, Arizona. *© AP/Press Association Images*

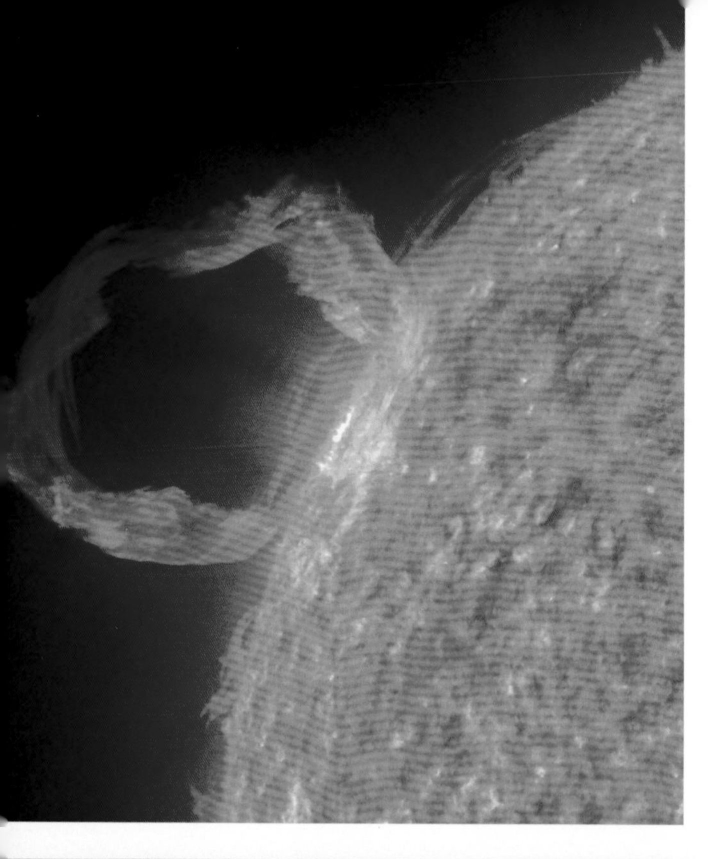

left: A fiery prominence larger than the Earth rears upwards from the Sun's incandescent surface. More powerful magnetic eruptions on the Sun have the potential to destroy our civilisation.

below: Crucibles of starbirth in Sagittarius: the Lagoon Nebula (bottom), and the Trifid Nebula (top). These beautiful clouds of glowing gas are lit up by hundreds of fledgling stars.

'he Crab Nebula is the guts of an exploded star, its skeins of gas rich in freshly created elements. 'he core has survived the supernova explosion, as a tiny spinning pulsar. *© NASA via Getty Images*

right: Thornhill Church, in Yorkshire's Calder Valley. It was here, in 1783, that the Reverend John Michell first postulated the existence of black holes.

© Hencoup Enterprises

left: In New Jersey, Arno Penzias (left), and Robert Wilson (right) stand in front of the horn antenna with which they discovered the microwave backgroun in 1965 – the afterglo of creation.

© Roger Ressmeyer/CORBI

right: Maverick Swiss astronomer Fritz Zwicky, working with his despised American colleagues in the 1930s, pioneered research on today's hot cosmic topics: the invisible forces of dark matter and dark energy.

© STR/AP/Press Association Images

above: The most spectacular comet of recent times, Comet McNaught. Visible in southern skies, it boasted a plethora of tails and was the biggest comet ever seen – outshining even Venus. *© S.Deiries/ESO*

left: The Rosetta space probe closes in on Comet Churyumov-Gerasimenko. As it heats up on approaching the Sun, the dirty snowball starts to eject jets of steam. Rosetta will orbit the comet for over a year. *© NASA Ames/SETI Institute/JPL-Caltech*

right: Blazing through the sky above Russia in 2013, a small asteroid ended its space career on Earth. The explosion severely damaged the city of Chelyaninsk – but miraculously nobody was killed. *© Ria Novosti/Science Photo Library*

left: Mars's spectacular volcanic region – Tharsis – in all its glory. The biggest volcano, Olympus Mons (centre) is three times higher than Mount Everest, and could cover the whole of Spain.
© Mark Garlick Words & Pictures Ltd/Science Photo Library/Corbis

below: Earth's biggest radio telescope: the Arecibo dish in the jungles of Puerto Rico measures 305 metres (1000 feet) across. It has been used to broadcast a message to a putativ extraterrestrial civilization.
© Stephanie Maze/CORB

its core is limited: one day, it will run out. Now it's time for gravity to resume its aeons-long battle with the star, which has been going on since its birth. Gravity shrinks the core again, giving it the chance to 'burn' helium – the product of the previous hydrogen reaction – into carbon and oxygen. This new, hotter process has a marked effect on the star's girth: its atmosphere billows out, like a distended balloon. The star has become a red giant.

It's a fate that awaits our Sun. As it swells up, our star will swallow up Mercury; then Venus. Slowly, inexorably, its incandescent surface swells outwards – towards planet Earth. As the Sun turns the heat up, the oceans boil away. As seen from Earth, the baleful Sun grows in size, until it stretches from horizon to horizon. Then its hot gases engulf the Earth. Our planet melts. And then the Earth evaporates, in a puff of vapour that becomes one with our parent Sun.

The red giant Sun will be huge – extending farther out into space than the Earth's orbit. But this pales into insignificance compared to some of the red bloated denizens of the cosmos, which are truly gargantuan.

In 1920, the American physicist Albert Michelson, chiefly famed for measuring the speed of light, used his mastery of optics to measure the diameter of the red giant Betelgeuse. Using the famous Hundred-Inch Telescope on California's Mount Wilson, he combined separate light beams from the star to ascertain its girth. Michelson discovered that Betelgeuse was literally hundreds of times wider than the Sun. It's a star that, if placed in the Solar System, would extend beyond the asteroid belt – swamping the planets all the way out to Jupiter.

As the news emerged, reporters flocked to Michelson's house. A scientist of the old school, Michelson had little time for the press: 'Reporters are sure to get the facts all wrong and to give the information some sensational twist to catch the public interest.' He refused to talk to the press, and his worst fears about sensationalism were consequently confirmed. 'Michelson Measures Colossus of the Skies', the headlines screamed. 'As for Our Little Earth, Betelgeuse is as Big as Trillions of Globes Like It!'

Red giants like Betelgeuse, and its rival Antares on the other side of the sky, find it hard to get to grips with themselves. Gravity has a weak hold on their outer layers. These colossi of the cosmos wobble like a blancmange, changing in brightness as their atmospheres billow in and out. Eventually, something has to give. The red giant becomes increasingly unstable, and – finally – the star gently puffs its distended atmosphere into space. The result is a glorious cosmic smoke-ring: a planetary nebula.

The astronomer responsible for coining the phrase was William Herschel, who – now based with King George III near Windsor – had turned his great telescopes towards the nebulae. Early in September 1782, he observed a peculiar object in Aquarius: 'A curious nebula, or what else to call it I do not know. It is of a shape somewhat oval, nearly circular …'

It reminded Herschel of the planet Uranus that he'd discovered – so he called it a 'planetary nebula', and the name stuck.

Gossamer-thin, planetary nebulae are the will-o'-the-wisps of the cosmos: they are astonishingly delicate and ephemeral. They are also very beautiful, and often complex. The Ring Nebula, the Helix, the Owl Nebula, the Dumbbell, the Cat's Eye Nebula – all are lovely examples, and fabulous photographic targets.

Planetary nebulae live for only tens of thousands of years before evaporating away into the darkness of space. But despite their short lifetimes, they are incredibly important for future starbirth. They seed the cosmos with elements that were built up in the dying star: atoms of carbon, nitrogen and oxygen that could form the basis of life on planets around stars yet to be born.

But the story of a star is not yet over. Peer through the shimmering veils of a planetary nebula into its centre, and you'll discover a tiny point of light. It's the star's former core – once a celestial powerhouse. Now, bereft of its nuclear reactions, it lies crushed and collapsed. This tiny object has the mass of the Sun, but is the size of the Earth. Astronomers called these cosmic zombies – half-living, half-dead – 'white dwarfs'.

The nearest white dwarf lies 8.6 light years away. The first astronomer to suspect its existence (although he couldn't have any idea as to its nature) was Friedrich Bessel. In the 1830s, he was measuring the positions of stars that he suspected lay near the Earth, scrutinising them for 'wobbles' that might reveal their distances.

He homed in on Sirius – the brightest star in the sky – and discovered that it was indeed wobbling, but over a period of decades: a timescale that was far too long. He concluded that it was being circled by an unseen companion.

Two decades later, the leading American telescope maker, Alvan Graham Clark, was testing out his latest instrument on Sirius. To his astonishment, he discovered that it was accompanied by a tiny star 10,000 times fainter. Because it keeps the 'Dog Star' company, astronomers nicknamed this fainter star 'The Pup'.

At first, no one was too surprised. After all, stars come in a range of brightnesses, and The Pup could have been a cool star, giving off less light than white-hot Sirius. But in 1914, astronomers discovered that the star was every bit as hot as Sirius itself.

If the two stars were equally hot, how could they shine with such different brightnesses? The only answer was that The Pup was very much smaller than Sirius: around the size of the Earth. This astonished astronomers – they had always assumed that stars had to be much bigger than planets.

The final shock was when astronomers checked the orbit of The Pup, and weighed it: it turned out to be as heavy as the Sun. An object like this was simply unbelievable. In the words of the great stellar astronomer Sir Arthur Eddington:

> We learn about the stars by receiving and interpreting the messages that their light brings to us. The message of the Companion of Sirius when it was decoded ran: 'I am composed of material 3,000 times denser than anything you have ever come across; a ton of my material would be a little nugget that you could put in a matchbox.' What reply can one make to such a message? The reply which most of us made in 1914 was – 'Shut up. Don't talk nonsense.'

Gravity – which dictated the very birth of a star – has triumphed. It may have taken billions of years, but now it has broken up the very fabric of the star that it created. The star's atoms have been crushed into a dense sea of nuclei and electrons. This is the end of the road for the vast majority of stars in our Galaxy.

Although a white dwarf starts off blazingly hot, the only way now is downhill. With no nuclear reactions to drive it, all it can do is to cool. Like the dying embers of a once-raging fire, the dwarf will dwindle from brilliant white, to yellow-orange, then dull red.

If we could revisit the Solar System in billions of years, we'd find merely a dim white dwarf, circled by a devastatingly culled family of planets. The innermost is a severely charred Mars. Jupiter and Saturn are shadows of their former selves, their gases boiled away during the Sun's red giant phase. And the white-dwarf Sun will continue to fade as it surrenders the last of its heat to space. Like other stars that have gone before it, the Sun will end its days as a cold, dark cinder – a forgotten piece of clinker in the cosmos.

Chapter 10

STARDOOM

It was 4.30 on a cold October morning in 1997. Alone in the dark dome of his backyard observatory, computer engineer Tom Boles was terrified ...

No, he wasn't worried about intruders, wild beasts or internet viruses invading his garden in the Midlands of England. It was the latest image from his telescope – displayed on a computer screen – that was making his heart pound.

'There was an extra bright area in the galaxy NGC 3451,' Boles recalls. This speck of light appeared in a galaxy – a giant star-city like our Milky Way – lying so far away that its light left NGC 3451 when dinosaurs were still walking the Earth. He realised that the intruder on his computer screen could be Boles's first discovery of a supernova – a distant star exploding in a cataclysmic fireball billions of times more brilliant than the Sun.

But Boles faced an immediate problem: the 'bright area' could be simply a star on our doorstep flaring up, or an asteroid in our Solar System that happened to be passing in front of the galaxy.

'I'd been looking for a discovery like this for over a year,' says Boles, 'but the truth is that – when it happens – you are not ready for it.'

If he was convinced he'd found a new supernova, he would contact the International Astronomical Union. They would alert the world's major observatories, where professional astronomers would swing some of the world's biggest telescopes towards the galaxy Boles had pinpointed.

'What if a large telescope were to be taken away from its scheduled research,' he explains, 'and I had made a mistake ... The terror deepened and deepened.'

Boles spent days checking and rechecking, along with a small team of British amateur astronomers. Eventually, they sent an email to the quaintly named Central Bureau for Astronomical Telegrams, in Cambridge, Massachusetts. Here, Brian Marsden – an ebullient Englishman – ran a clearing house for all new astronomical discoveries on behalf of the IAU. Marsden would check any reported new discoveries, then inform the world's major observatories. Despite the organisation's name, these alerts are now sent electronically, rather than by telegram!

'The tough part was yet to come; a long period of silence followed,' Boles tells us. 'Four days passed. Those were the longest four days of my life. Doubts started to set in ...'

Then the news arrived, in the IAU's Circular No. 6763. It reported 'the discovery of an apparent supernova (magnitude about 16.0) by Tom Boles, Wellingborough'. Followed by the clincher. Professional astronomers had used a large telescope in Arizona to split the light of the suspect object into a spectrum, which proved it was indeed 'a type-II supernova after maximum'.

'If it was confirmed, I'd expected to feel elation,' Boles confides. 'No – I was relieved and tired. The night after the telegram arrived, I slept for fourteen hours.'

The next day, Boles received a personal message from Marsden. It simply read: 'Well done! Just to prove that this wasn't a fluke, find another one.'

And he did. By 2014, Boles's total had reached a staggering 155 supernovae – more than any other astronomer in history, amateur or professional. And he's still going strong!

How does it feel to be the world's most prolific supernova hunter? Boles is typically modest: 'I got used to that feeling some time ago, and I don't think too much about it now,' he says. In particular, he can now contact the professional astronomers directly: 'My reputation does help me to get attention!'

Tom Boles follows in the footsteps of a legendary supernova-hunter from the other side of the world. The Reverend Robert Evans, who lives in New South Wales, has two heavenly interests: by day he preaches in the Uniting Church; by night he scours distant galaxies for supernovae.

Both Boles and Evans observe through medium-sized backyard telescopes, but there's one crucial difference. While Boles uses an electronic camera linked to a computer, Evans employs much simpler equipment: his eyeball.

Since his first supernova discovery in 1981, Evans has built up a list of 1,500 galaxies to scrutinise. 'I have tried to memorise the appearance of many of these galaxies,' he explains, 'so that I can recognise instantly any new feature.' He plays down this remarkable memory feat. 'Most people can remember pretty well hundreds and hundreds of different faces of their friends and relatives. These galaxies are my friends.'

If a supernova erupts in one of his galaxies, it leaps out at Bob Evans like an angry pimple on a best friend's cheek. He holds the all-time record for discovering supernovae using just the human eye as the telescope – a phenomenal forty-two exploding stars.

But clouds over his home in the Blue Mountains, one night in February 1987, frustrated Evans from a discovery that would have been the ultimate jewel in his supernova crown.

The skies were clear, however, at the Las Campañas observatory in Chile. Here, young Canadian astronomer Ian Shelton had been taking regular photographs of the Large Magellanic Cloud. As he pulled a glass negative from the dish of chemical developers, in the early hours of 24 February, 'I realised there was something very strange about the plate. There was one extra bright star on it.'

The Large Magellanic Cloud appears as a glowing patch in the skies of the southern hemisphere, like a portion of the Milky Way that's been torn off and flung onto the black sky nearby. It's prompted many myths. In Western Australia, the Aboriginal people see the Large Magellanic Cloud as the campsite of an old man who's too frail to catch his own food; the nearby stars bring him fish from the celestial river marked out by the Milky Way.

Less romantically, we know today that the Large Magellanic Cloud is a swarm of stars – the nearest galaxy to the Milky Way Galaxy where we live. Even so, 'nearest' means a distance of 160,000 light years – so we are viewing this galaxy as it was about the time when modern humans – *Homo sapiens* – first evolved in Africa.

Still dazed by the strange object in the photograph, Shelton rushed outside and looked up at the glowing Large Magellanic Cloud. He could see the new star with his unaided eyes 'as plain as day; well, plain as night!' It was so brilliant it had to be a supernova – the closest observed in almost 400 years.

Shelton hurried to a neighbouring telescope on the mountain-top observatory, and alerted the astronomers there. 'We just all got together, walked outside, and stared at it. It's quite miraculous.'

They tried to phone their discovery through to the Central Bureau for Astronomical Telegrams, but in the days before email they – frustratingly – were just put onto voicemail. They couldn't alert the world's astronomers to the supernova; and pretty soon someone else was going to notice the brilliant object in the Large Magellanic Cloud. 'We just got desperate; the Sun was already up,' says Shelton. 'And we thought, pretty soon it's getting dark over the other side of the world.'

And – as night fell over New Zealand – Albert Jones settled in with his telescope. A retired miller by profession, Jones spent all his spare time checking out stars that vary in brightness. By the time of his death in 2013, Jones had made an astounding 515,000 observations of variable stars – far ahead of any other astronomer in history. It was such an amazing achievement that the Queen awarded him one of the British Commonwealth's top honours, the Order of the British Empire.

That night in February 1987, Jones turned his telescope towards the Large Magellanic Cloud, to explore its treasure-house of variable stars. 'And there was something in the field of view that wasn't there the night before,' he told us. 'Goodness gracious me, I said to myself. This is unusual – a bright blue star.' Jones measured the star's brightness; then he called up a colleague, 'who phoned the news through to the big observatory at Siding Spring in Australia.'

High in the Warrumbungle Mountains, Scottish astronomer Rob McNaught received the call. 'I rushed out in a state of shock and excitement,' he told us soon after the event. 'Sure enough, there was a bright supernova in the Large Magellanic Cloud.'

McNaught had to hand a large camera that he usually employed to track satellites crossing the night sky. He immediately swung it to the Large Magellanic Cloud, to photograph the intruder. Crucially, he could pinpoint its position precisely. 'And I looked at old photographs and – sure enough – there was a faint star in just the same position as the supernova.'

McNaught had not just captured the exploding star – now known as Supernova 1987A – he'd found old pictures of how the star looked *before* it committed spectacular suicide. It was a heavyweight – twenty times heavier than our Sun – which had swollen to become a giant star. For the first time, astronomers had a real astronomical beast to check out against their calculations of stardeath.

'Supernovae were recognised as one of the most exotic phenomena in the universe only fifty years ago,' he added. 'So this was the first bright supernova that actually allowed us to study it in great detail. What happened till this time had been largely theory.'

That theory takes us back to the inside story of what makes a star tick. At the start of their lives, stars blossom into life by converting hydrogen into helium deep in their incandescent centres. The hydrogen fuel eventually runs out. The core shrinks. It grows hotter. And then helium suddenly ignites. It 'burns' into carbon; and the renewed nuclear fires make the star expand grossly, to become a huge distended red giant.

For a bantamweight like the Sun, this is the beginning of the end. As we saw in the last chapter, its outer layers puff off into space – as the beautiful cosmic smoke-ring of a planetary nebula – while the exposed core fades away as a planet-sized star, called a white dwarf.

But a heavyweight star is only just beginning to flex its muscles. Something exotic is happening in the middle of the distended giant star. Gravity squeezes its core to ever higher temperatures. In this nuclear inferno, carbon begins to burn to heavier elements.

It's all done in an orderly fashion, like a set of nested Russian dolls – or the shells inside an onion. If we delved downwards into a heavyweight star, we'd find a shell of carbon, then a shell of neon and a shell of silicon. Eventually, the nuclear fires in the very centre begin to burn silicon to iron. And now the star is in real trouble.

Iron is the most stable of all the elements. In nuclear terms, it's just ash: however much you stoke up the temperature, it's never going to burn into anything else. In fact, the increasing heat and pressure begin to smash up the iron ash. In a matter of seconds, the star's core implodes. Without any support, the inner layers of the stellar onion crash downwards.

And it gets worse. In these hellfires, the iron breaks down to tiny particles called neutrons, which form a small solid ball. 'The core that's imploding on itself suddenly comes to an abrupt stop,' says Ian Shelton. 'It was about as big as the Earth, and now it's just the size of a city.'

The plummeting gas all around smashes into this solid ball. It's like a powerful ocean wave hitting a sea wall: just as spume from the breaking wave is thrown high into the air, the star's infalling gas bounces outwards again at high speed, ripping the star apart.

That's the theory; and that's exactly what Ian Shelton and Albert Jones witnessed that night in 1987 – a shockwave ripping outwards from the star's heart, erupting in a tumult of flame.

The world's biggest telescopes tracked the expanding fireball. Decades later, the Hubble Space Telescope is still monitoring the ring of now-faint gas as it speeds away from the site of the dead star.

And there's one more telescope that proved the theory of exploding stars was spot-on. This instrument wasn't in space, nor even on a mountain top. It was deep underground, down a mine in Japan. The Kamiokande 'telescope' was a huge tank of ultra-pure water, monitored by exquisitely sensitive detectors that could pick out the tiniest glimmer of light in its stygian depths.

At 41 seconds past 7.35 on the morning of 23 February 1987, the tank sparked into life. The detectors saw eleven flashes of light in the giant water tank. A bevy of energetic particles had plunged unhindered through the Earth's solid bulk, to light up the dark Japanese tank. Only one kind of particle could produce these subterranean fireworks: a ghostly neutrino, that can penetrate through virtually any thickness of matter.

And only the most infernal places in the universe can create neutrinos. High up on the list is the core of a supernova explosion, as the iron core collapses to a ball of neutrons.

'All hell is breaking loose at the centre,' explains Shelton. The temperature soars to around 50 billion degrees Celsius – thousands of times hotter than the centre of our Sun. It's the ideal place to make neutrinos.

Neutrinos from the core of Supernova 1987A had winged their way through space and lit up the Japanese tank. Though less than a dozen were detected, these exotic particles carried an amazing message from the centre of the dying star. For even this small number to reach the Earth, the original flood of neutrinos must have carried away 99 per cent of the supernova's energy. The neutrinos ripped the star apart in a colossal explosion that took a day to mature into the brilliant 'new star' that Shelton and Jones witnessed the following morning; but even this cosmic fireball was only a very minor sideshow to the neutrino outburst.

'The energy released at that instant,' enthuses Shelton, 'probably outshone – for a few moments – everything else in the universe.'

After this blaze of glory, a supernova's tiny core is left as a mere speck on the vast cosmic landscape: a tiny dense globe that theorists had dubbed a 'neutron star'. No one expected astronomers could actually discover such a minuscule star-corpse far out in space.

Neutron stars were certainly far from the mind of young researcher Jocelyn Bell when she scanned the sky with a bizarre-looking radio telescope at Cambridge, as the winter of 1967 closed in.

'We'd spent two years' hard labour on this new telescope,' recalls Bell (now Jocelyn Bell Burnell). 'It was like two thousand TV aerials covering four acres. They had to be held up out of the wet grass on wooden posts – so there was a lot of sledge-hammering and soldering and brazing and welding.' The grass under the telescope – covering the area of sixty tennis courts – was kept short by grazing sheep.

The great day came when Bell began to analyse her scans of the sky. The natural radio waves from space were traced out by a pen recorder – just as an electrocardiogram (ECG) displays your heartbeat in a continuous wiggly line. Bell wasn't looking for a *regular* beat from space, though: she was on the search for radio signals that fluctuated wildly, like the twinkling of the stars we see at night.

'But, as the pen ran over the chart paper,' Bell explains, 'it went beep, beep, beep – regular pulses about one and a third seconds apart.' To Bell's utter amazement, the pen recorder was displaying a steady pulse.

Most likely, it seemed, it was just interference from some badly suppressed electrical equipment nearby. Her supervisor Tony Hewish – who had designed the telescope – dug deep into the signal. To his own surprise, he had to conclude the radio waves were coming from something way out into space, hundreds or thousands of light years from us.

'It looked so regular,' says Hewish, 'that we made a joke and we called this thing "Little Green Men" – or LGM.' The head of the radio astronomy observatory, Martin Ryle, was alarmed. Suppose these were actually broadcasts from Little Green Men – and the news got out – then someone on Earth might invite the aliens to visit us. Ryle was convinced they would destroy our civilisation. If the signals turned out to be an intelligent message, Ryle decreed, his team must destroy all their records.

Bell had a more laid-back attitude. She had set out to study twinkling radio sources in the universe – but 'some silly lot of green men had to choose my aerial to communicate with us!'

As everyone else celebrated the approach of Christmas, Bell was immersed in endless rolls of chart recordings. Another suspicious signal caught her eye, and she headed out to the observatory to check it out. On her Lambretta scooter, Bell had to face a five-mile drive into the teeth of the fenland winter winds.

'The telescope was in a half-dead state because of the cold,' she says, 'but I breathed on it – and swore – and got it working. And as the chart paper flowed under the pen, the pen went beep, beep, beep. This was terrific! It was highly unlikely that two lots of little green men on opposite sides of the universe were both signalling to the Earth.'

Bell and Hewish had found not alien life, but a new inhabitant of the galactic zoo. 'We called them "pulsating radio sources",' says Hewish. But the boring scientific term was soon usurped. 'The word "pulsar" was, I believe, coined by the science correspondent of the *Daily Telegraph*.'

The discovery wasn't just a matter of pure luck. The Cambridge team had diverted from their original plan to investigate the strange signal. In contrast, an American radio astronomer a few years earlier had been surveying the sky when the pen of his chart recorder started dashing back-and-forth across the paper. 'Oscillation in the receiver,' he muttered, and fixed the problem – as we all do – by kicking his equipment. The pen stopped. He missed the discovery; and lost out on a Nobel Prize!

But what *was* a pulsar? It had to be incredibly powerful. And its rapid pulses meant it was a little beast – smaller than the Sun, and even tinier than the most petite star then known, a white dwarf. The only possibility was a neutron star – the collapsed core of a supernova.

It may be small, but a neutron star packs a mighty punch. It's only the size of London or New York, but contains as much matter as a million Earths. Scoop a pen-cap of material from a pulsar, and it would weigh a million tonnes.

'If you could land successfully on a neutron star,' says Bell Burnell, 'you'd experience phenomenal gravity. Suppose you try to climb a little mountain – say a centimetre high – you'd have to do as much work as climbing Mount Everest on Earth.'

As well as its immense gravity, a neutron star punches well above its weight when it comes to magnetism. The tiny beast has a magnetic field a trillion times stronger than the Earth's. The pulsar has two magnetic poles – like our planet – and here the concentrated energy sends powerful beams of radio waves into space.

'The beam is swept around as the neutron star rotates,' says Bell Burnell. 'It's like a lighthouse. When the beam sweeps across us, we pick up a pulse of radio waves.'

So pulsars are cosmic corpses that, like zombies, have a life after death. But even pulsars don't wield the ultimate in unseen cosmic powers. More recently, astronomers have discovered magnetars – neutron stars with magnetism that's a hundred times more powerful still. If a magnetar passed our planet at the distance of the Moon, its magnetism would wipe the information from every credit card on Earth!

England's Astronomer Royal, Sir Martin Rees, is captivated by these masters of the cosmos. 'Magnetars are of special interest to physicists,' he explains, 'because they allow us to extend our knowledge of the basic laws of physics to breaking point.'

Most of the neutron stars – pulsars and magnetars – in our Galaxy were born in supernovae that erupted millions of years ago. As time goes by, the cosmic lighthouse slows down, its pulse rate declines and it fades from view.

But there's one that stands out from the crowd. The Crab Pulsar had its fiery baptism in a supernova that erupted less than a thousand years ago – as the ancient Chinese astronomer Yang Wei-te related: 'I have observed the appearance of a guest star, in the 5th moon in the eastern heaven in T'iem-kuan. It was visible by day, like Venus; pointed rays shot out from it on all sides.'

By our calendar, it was AD 1054; and the brilliant 'guest star' blazed forth in the constellation Taurus. This supernova shone in daylight for twenty-three days; and was visible at night for two years.

Knowing nothing of the Chinese records, an English physician – and amateur astronomer – John Bevis noted a small shining cloud of gas, when observing this part of the heavens in 1731. Sadly, his charts showing the enigmatic object were never widely distributed, as his publisher went bankrupt. Bevis himself died after falling from his telescope at the age of seventy-six.

Soon after Bevis's sad demise, French astronomer Charles Messier included his nebula as the first of a list of fuzzy celestial objects that could be confused with comets – Messier's chief passion. A century later, it fell under the scrutiny of the Third Earl of Rosse, the wealthy amateur astronomer who'd built a major observatory in the middle of the Irish bogs.

'It is no longer an oval resolvable Nebula,' Rosse noted. 'We see resolvable filaments singularly disposed.' His assistant wrote of 'streamers running out like claws in every direction' – which is probably why it became known as the Crab Nebula.

By the 1960s, astronomers had put together the Chinese and modern records, and worked out that the Crab Nebula was the fireball of a supernova that had blown up some 900 years earlier. At its heart was a strange blue star – but, before the discovery of pulsars, no one knew quite how strange. A great scientific discovery was missed, however, when professional astronomers refused to listen to a member of the public.

'In the late 1950s,' says Jocelyn Bell Burnell, 'an observatory in Chicago had an open night to show the public the stars. The telescope was set to a peculiar star near the centre of the Crab Nebula. One woman looked and said, "That star's flashing!" – but no one believed her.'

The heart of the Crab Nebula is indeed marked by a pulsar that's flashing thirty times every second, right on the borderline of what the human eye can detect. The unknown visitor had probably eyeballed a pulsar, a decade before Bell and Hewish made their epic discovery. The flashes of light from the Crab Pulsar were only detected much later, when a British team observed the region with a large telescope and an electronic camera. Their comment – recorded on tape – says it all: 'It's bloody pulsing!'

By the 1960s, the Crab Nebula had become an iconic object. British astrophysicist Geoffrey Burbidge quipped: 'You can divide astronomy into two parts: the astronomy of the Crab Nebula and the astronomy of everything else!'

The Crab Nebula was close to Burbidge's heart. It could contain the proof of something he'd spent years working on: how supernovae produce new elements.

It was a team effort, involving four of the greatest minds of the twentieth century. Margaret Burbidge – Geoffrey's wife – was a leading celestial sleuth, decoding the light from stars and nebulae to prise out what they were made of. Willy Fowler worked in a physics lab, investigating the ins and outs of nuclear reactions. Bringing them all together was the brilliant astrophysicist Fred Hoyle.

'Fred was the founding genius,' recalled Geoffrey Burbidge. 'Willy was the experimental nuclear physicist. Margaret was the observer; and I could bring a lot to it in terms of creative physics.'

The formidable team of Burbidge, Burbidge, Fowler and Hoyle was universally known as B^2FH (pronounced B-squared-F-H). In the 1950s, they had predicted that stars would explode when their cores became clogged with iron 'ash'. And B^2FH also calculated just how the nuclear furnaces inside a star fulfilled the old alchemists' dream of converting one element into another.

In the beginning, the universe was made of only hydrogen and helium. Everything else was forged inside stars. The Crab Nebula's 'claws' are rich with atoms that have been made inside the old star – nitrogen, sulphur and argon. The glowing nebula is literally the guts of the old star, which have been spewed out into space.

And when astronomers have investigated the gas clouds thrown out from other dead stars, they've found an even richer cornucopia of newly made elements.

So the legacy of a long-dead star lives on. The fireball from the explosion smashes into surrounding clouds of gas, squeezing them so tight that they condense into a new generation of stars – and accompanying planets. And a dying star bequeaths a rich bounty of freshly minted atoms to these new worlds. Our planet Earth was created from the ashes of an old supernova. The carbon atoms in our bodies – and the gold in your wedding ring – were forged inside stars that died long ago.

At the end of the day, we shouldn't mourn the death of a star as a supernova. Because a star is a phoenix: from its ashes arises a whole new generation of richly endowed stars and planets.

Chapter 11

BLACK HOLES

'This is an object that is not made from ordinary matter,' muses physicist Kip Thorne. 'It's not made of antimatter. It isn't made of matter at all. It's made out of a pure warpage of space and time.'

'Nothing comes out of it; things only go into it,' continues astronomer Phil Charles. 'It's that one-way nature which probably upsets people most, causes them to be really disturbed – and, in fact, causes many scientists to be disturbed.'

Welcome to the sci-fi world of black holes: objects which bend the very fabric of space, defy the laws of physics, and may even be pathways to other universes.

But for all their exotic contemporary connotations, black holes had a surprisingly parochial, genteel – and English – beginning. Visit the village of Thornhill in Yorkshire, and you'll discover a delightful mediaeval church perched above the valley of the River Calder. In the eighteenth century, it was home to rector John Michell and his family.

Very little is known about the man. There are no portraits; no letters; no personal papers. His memorial in the church reads that he was: 'Twenty-six years rector of this parish, eminently distinguished as a philosopher and scholar. He had a just claim to the character of a real Christian, in the relative and social duties of life, a tender husband, the indulgent parent, the affectionate brother and the sincere friend were prominent features in a character uniformly amiable.'

We know that Michell was educated at Cambridge, and was certainly a polymath, covering the fields of maths, Greek, Hebrew and theology. But his strongest point was geology. After the Lisbon earthquake of 1755, he published a paper in which he correctly deduced that earthquakes were carried by waves under the Earth's surface. For this, he was awarded fellowship of London's prestigious Royal Society.

In 1783, he had been contemplating his other burning passion: astronomy. Michell had been invited to read a paper to the Royal Society, and decided to address the nature of gravity in stars. How, he wondered, did gravity affect the light flooding out of a star's surface?

Michell reasoned that a very massive star might pull back its own light, just as a ball thrown upwards will fall back to Earth. Then he came up with his masterstroke: 'All light emitted from such a body would be made to return towards it, by its own proper gravity ... if there should really exist in nature any (such) bodies, their light could not arrive at us.'

Astronomer Royal Martin Rees is full of admiration for Michell, who was centuries ahead of his time. 'John Michell worked out that if you had a body weighing about a hundred million times as much as the Sun, that light couldn't escape from it. And he went on to say that – for that reason – maybe the most massive objects in the universe might be invisible to us.'

In short, an eighteenth-century English rector had predicted the existence of black holes.

A star 100 million times heavier than the Sun is a physical impossibility. The most massive stars known outweigh the Sun 200 times over – but even that's uncertain, because of the difficulties in measuring the properties of stars so far away. But gravity has a trick up its sleeve. Squeeze a star to an incredibly high density, and its pull will become irresistible.

If you could shrink the Earth down to the size of a large marble, its gravity would be strong enough to prevent light from escaping. Fortunately, there's no chance of that happening! But the death throes of stars do produce ultra-compressed corpses, with correspondingly powerful gravity. When the Sun dies, its core will collapse to become a white dwarf star: the same size as the Earth, but containing 200,000 times as much matter. To escape from a white dwarf, you would need to travel at more than 6,000 kilometres per second.

Even smaller and denser are neutron stars, the ultra-compressed cosmic zombies forged in the heat and power of a supernova explosion: they're often called pulsars, because they send out pulses of radio waves as they spin around. The escape velocity for a neutron star is a staggering 96,000 kilometres per second: almost one-third of the speed of light.

After Jocelyn Bell and Tony Hewish discovered pulsars in 1967, astronomers got on the scent of an even more exotic quarry: an object whose escape velocity *was* the speed of light.

Martin Rees explains: 'We believe that many black holes are created as the end point of heavy stars.' Most of these exploding stars – supernovae – leave a pulsar or neutron star as their bequest to the universe. 'But in some of these supernovae,' continues Rees, 'probably those that result from a star that is very heavy – twenty or thirty times as much as the Sun – the remnant may instead be a black hole.'

Back to the drawing board. Astronomers calculated that neutron stars have a natural weight limit. Even neutrons, standing shoulder-to-shoulder, ultimately can't counter the pull of gravity. Any supernova relic weighing in at more than three times the mass of our Sun is destined to collapse for ever. The result will be a black hole.

In the story of the universe, gravity is king. It is the starmaker – and, ultimately, the starbreaker.

Why 'black hole'? They are named after the 'black hole of Calcutta': a cell in an eighteenth-century Indian fort normally used to hold three prisoners. On one occasion, 146 unfortunates were crammed into the space – and only 23 of them came out alive …

A black hole is the ultimate cosmic abyss. The corpse of a supermassive star is black, because not even light – travelling at 300,000 kilometres per second – can escape its mighty gravity. And it's a hole, because anything that falls in is trapped there forever. The reason is simple: the velocity of light is the speed limit of the universe – if light can't escape, nor can anything else. Black holes are the graveyards of the cosmos. They are places where matter disappears from sight, and can never return.

So how do you detect a black hole – an ebony entity in the darkness of space? It helps if the black hole is having an altercation with another star. Think of a pair of cats in a coal-cellar, having a spat at night. One is white; the other is black. You won't be able to see the black cat, but you'll certainly notice from the behaviour of its white adversary that something pretty ferocious is going on!

But to be party to the giant cosmic cat-fight, you have to develop a new technology. Cue X-ray astronomy. Confusingly, it has nothing to do with looking at the innards of celestial beings. Instead, it focuses on the incredible energy some objects emit.

Think of natural radiation from the sky as the notes on a piano. The light that we see comprises the notes around middle-C. Some cosmic entities emit low-frequency bass notes like radio waves; frenetic bodies such as black holes sing out celestial tremolos at the highest end of the wavelength register. Alas: these cosmic countertenors can't be heard through our thick and murky atmosphere. To detect the song of the black hole, you must send your X-ray detector into space.

And so it was that, on 12 December 1970, American scientists launched a revolutionary satellite to scan the universe for X-rays. The launch site was Kenya, and the date was the seventh anniversary of the country's independence: so the satellite was christened Uhuru – Swahili for 'freedom'.

Uhuru was an incredible success. It discovered hundreds of X-ray sources: most of them neutron stars ripping gas from their companion star. But Cygnus X-1 was different. Astronomers pinned it down to a hot, blue star weighing in at some thirty Suns. But it was being circled by an unseen object ten times more massive than our local star – far too heavy to be a neutron star. It had to be a black hole.

The black hole is seizing matter from its blue giant companion – to spectacular effect. The superheated gas streams towards the black hole, where it forms a fearsome cosmic maelstrom – the accretion disc – that gives out a scream of powerful X-rays before disappearing from our universe forever.

Cygnus X-1 was the first black hole to be discovered – but it's far from being the only one. This cosmic beast has inspired a new breed of astronomer: the black hole hunter. And Phil Charles leads the pack.

'To be a successful black hole hunter,' he explains, 'you have to have ready access to the largest ground-based telescopes.' It's the search for the darkest objects in the universe – rather than tropical sunshine – which tempts Charles to La Palma, or Hawaii or Australia, the location of some of the world's best telescopes. He pinpoints stars where astronomers have discovered a loud shriek of X-rays, the signature of the elusive black hole: 'When you're looking for a needle in a haystack, we need the needle to shout out and say "I'm here".' Charles then tracks the motion of the visible star as it orbits its unseen companion, which gives away the weight of the black hole.

Thanks to Charles and his colleagues, astronomers have now pinned down a score of black holes in the Milky Way, ranging in mass from four times more massive than the Sun up to sixteen times heavier. But these are just the tip of the iceberg: according to Martin Rees, 'there's every reason to suspect that in our Galaxy there are many millions of black holes weighing perhaps ten or twenty times as much as the Sun'.

That means there could be one black hole for every thousand stars we see in the sky. And the links between black holes and stars run much deeper than merely sharing the same Galaxy. By a superb cosmic irony, the darkest objects in the Milky Way were born from the death of the very brightest and most massive stars which explode as supernovae.

But why are black holes so controversial? It took a long time for many conservative astronomers to accept that they existed. As late as 1970, the distinguished astronomer George McVittie – at an international conference – scrawled on the old-fashioned blackboard: 'Black Hole = Naughty Word'.

The answer lies with Albert Einstein (who was actually unhappy with the idea of black holes himself). His new theory of gravity, published in 1916 – general relativity – takes a different tack from that of Isaac Newton, who saw gravity as a mysterious force that reached out across space.

Einstein concentrated his efforts on gravity at its most powerful. And, in the end, he concluded that its effects were profound. A celestial body warps the very fabric of time and space, like a ball-bearing denting a rubber sheet. In Einstein's gravity, nearby objects roll into the hollow – but there are bizarre effects as well.

In the case of extreme gravity, Einstein's equations predict that objects falling towards a powerful attractor will end up travelling at the speed of light. The consequences are awesome: the infalling object will acquire infinite mass; time will appear to stop for it; and the object itself will be 'spaghettified' into a long, thin tube before it meets its demise.

So what actually happens when something falls *inside* a black hole? Here, we're in the realms of pure calculation. After all, if no light – and therefore no information – can emerge, how can we know what's going on? You need to be a maths wizard to do the sums – and then not everyone agrees.

The consensus is that the gases snatched by the black hole end up at its centre as a pinpoint of infinite density: a singularity. According to black hole guru Kip Thorne: 'The singularity is a place where gravity is infinitely strong, where gravity destroys matter, it destroys space, and it destroys time.' And he happily adds an observation as to what would happen if an astronaut were to fall into a black hole: 'The singularity destroys all the atoms of your body, transmuting them into the form of this singularity.'

Other physicists are a tad more optimistic. If the black hole is spinning at breakneck speed, then the dire effects of gravity could be overcome by the forces of rotation. In which case, there's a slim chance that a rotating black hole could be a gateway into another universe.

It's no wonder that astronomers have been confused about black holes.

A black hole weighing in at ten times heavier than the Sun, like Cygnus X-1, is a mighty cosmic proposition. But imagine the havoc a black hole that's a million – or even a billion – times more massive could wreak. It's these supermonsters that we turn to next …

We're heading for the centre of our Galaxy. Lying over 25,000 light years away, among the dense star clouds of Sagittarius, it is not the easiest part of the Milky Way to access. In fact, it's invisible to even the most powerful telescopes. Dark dust grains – cosmic soot – pile up along the way, completely blocking our view of the galactic centre.

Our Galaxy is shaped rather like a fried egg. The flat 'white' is home to the spiral arms, which cradle active, hot young blue stars. The bulging central 'yolk' is more sedate, made up of ageing red stars that are past their best: a cosmic retirement community. So – no chance of cosmic fireworks in there, astronomers once thought.

They were wrong. In the 1950s, radio astronomers started to discover suspicious signs of discontent among our Galaxy's elderly residents. Newly developed radio telescopes had one great advantage over conventional optical telescopes. Tuning into longer wavelengths, they could penetrate the murky dust that clogs up our view of the galactic centre. To these giant leviathans of engineering, the tiny dust grains were virtually transparent. It meant that they could home in on what was going on downtown in the Milky Way.

Soon they were seeing signs of rebellion in the old folks' home: expanding rings of gas; arcs; strong magnetic fields; high turbulence. Clear evidence that something was – or is – going on.

Infrared astronomers also had the power to peer through the galactic murk. They used giant telescopes, like those perched under the transparent skies atop Hawaii's Mauna Kea, to put the centre of our Milky Way under the microscope. They focused in on the stars circling our Galaxy's core. And they were astonished to find that these centre-stage stars were whirling around the heart of our Galaxy at astonishing speeds: 1,400 kilometres per second, which is 0.5 per cent of the speed of light. There could be only one culprit: something exerting very powerful gravity.

Radio astronomers then investigated the nature of a mysterious radio source which lay bang in the centre of the Galaxy. Called Sagittarius A*, this tiny source was emitting radio waves in profusion. From the orbits of the stars around it, they could now calculate how heavy it was – and it weighed in at over four *million* times the mass of our Sun.

In 2008, astronomers linked up radio telescopes in Hawaii, California and Arizona to pin down the size of this mysterious cosmic beast. It turned out to be just

44 million kilometres in diameter – less than the size of Mercury's orbit about the Sun.

Sagittarius A* had to be a supermassive black hole. That's the only way you can concentrate such prodigious amounts of matter into so small a region. What the radio telescopes were picking up was radiation from its accretion disc: the whirling disc of matter around the black hole proper. It's a vortex of gas from stars, torn-up stellar remnants and remains of planets that spirals into the hole, heating up to colossal temperatures as it does so.

Even so, the galactic centre's black hole is currently on a severe diet: there isn't enough material in our Galaxy's heart to satisfy its insatiable hunger. At the moment, it is in hibernation and quietly slumbering. But the gas rings and arcs first detected by radio astronomers are evidence for 'burps' from a violently active black hole in the past.

How common are supermassive black holes in galaxies? Kip Thorne observes: 'Lurking in the core of our Galaxy, and in the cores of most other galaxies, there appear to be gigantic black holes. They weigh between one hundred thousand and one thousand million times what the Sun weighs.'

Martin Rees adds: 'We believe that these form at about the same time the galaxy forms, from gas that settles into the centre. It passes the point of no return, where it can't form into stars, but has to contract as a single cloud. The cloud becomes a sort of "superstar", and then collapses to form a supermassive black hole.'

Look to the northern skies on a dark, clear autumn night, and you'll spot the Andromeda Galaxy: our closest major galactic neighbour. Some 2.5 million light years away, it's a spiral galaxy like the Milky Way, and a little bit larger. At its heart, it also plays host to a supermassive black hole estimated to weigh in at 100–200 million Suns.

But the black holes at the cores of the Milky Way and the Andromeda Galaxy are tiddlers on the cosmic scale – as astronomers were about to find out …

In the early 1960s, most astronomers regarded the universe as a fairly tame place. Placid stars whirled languidly around in picturesque galaxies, out as far as telescopes could see. At this time, the violent Big Bang was just a theory; other manifestations of the violent universe – pulsars, neutron stars and black holes – were just constructs in the minds of other-worldly theoreticians.

But all this was set to change … Radio astronomers in Cambridge scanned the heavens in detail with a newly built radio telescope. It was their third survey: hence '3C' for 'third Cambridge survey'. Among the objects they garnered were the usual suspects: the remains of exploded stars, and active galaxies. But there were a few that raised question marks, like 3C 273 and 3C 48.

These radio sources were an enigma. They looked like stars, but they were emitting powerful radio waves – which stars don't. Astronomers initially called them 'quasi-stellar radio objects', but this was mercifully shortened to 'quasars'.

What were they? Astronomers turned to optical telescopes to check out the spectrum of these 'radio stars', in order to find out the composition of the elements in their spectral lines. But their spectra didn't correspond to the lines in ordinary starlight.

Enter Maarten Schmidt – a young Dutch-American astronomer. In 1962, he used the great Two-Hundred-

Inch Telescope on Palomar Mountain to look at the spectra of both 3C 273 and 3C 48. These were the days before astronomers nursed their coffee in a warm control room, interacting with the telescope via a computer. Instead, the observer sat for many hours in the 'prime focus cage' inside the telescope – guiding the instrument in freezing conditions. What was it like? What do you do to while away the time and prevent nodding off?

Maarten Schmidt remembers the times rather fondly. 'I was really quite isolated. There were practical things to do, like guiding the telescope, looking for clouds, and making sure things went all right. It was a very romantic way of observing, really. Today it has developed into looking at a monitor.' Schmidt recalls feeling in touch with the stars – and listening to music. 'You had a long time to think – there was nothing else you could do. Music? BBC all the time, always classical music – a lot of Bach.'

Schmidt took his spectra back to the California Institute of Technology, and spent many a day scratching his head. There was something irksomely familiar in the pattern of the spectral lines of 3C 273 – until the penny finally dropped.

'I was irritated', he recalls. 'I said, damn it – I'll take my slide rule and I'll show that these lines are getting regularly spaced.'

Then it hit him.

'Jesus, a redshift. And immediately it struck me how incredible it was.'

Split up the light from any astronomical object, and you'll find it's actually a rainbow of colours – from long-wavelength red light, to short-wavelength blue. This spectrum is crisscrossed by vertical lines: a fingerprint of the different atoms that make up the star, planet or galaxy. These lines also reveal how fast a body is travelling through space. The expanding universe drags out the wavelengths of light from a galaxy towards the red – just as the pitch of a siren from a receding ambulance drops as it passes you. The greater this 'redshift', the further away the galaxy lies.

The redshift of 3C 273 was 16 per cent – corresponding to a distance of almost 2.5 billion light years.

Schmidt was astonished. The object wasn't a star in our own Galaxy. It actually lay further away than any galaxy known at the time; and to be so bright – at such a distance – it had to be forty times more luminous than a normal galaxy. Yet it was no bigger than the Solar System.

Schmidt remembers: 'That was something to think about. Anyway, in the ruckus, I made some noises, and my colleague Jesse Greenstein came by.'

Greenstein had been working on the spectrum of Schmidt's other quasar, 3C 48.

'He said, "My God", and produced his paper on 3C 48. Within fifteen minutes, we found out that its redshift was 37 per cent.'

Which meant that the object was nearly four billion light years away. It had to be a galaxy: but with something sinister lurking in its centre. Astronomers eventually realised that, once again, they were looking at the antics of black holes: but this time, writ very large. The black hole driving 3C 273 weighs in at a *billion* Suns. The mighty cosmic beast commands an accretion disc that screams with outbursts as it gobbles its matter at the rate of one Sun a year.

This solar-scale consumption isn't a smooth process. A quasar like 3C 273 gorges its food messily. Although most of the fodder goes down the throat of the black hole, the remains erupt in the accretion disc as a colossal cosmic burp. And these burps are formidable. In the case of 3C 273, they fuel a powerful jet of electrons – 200,000 light years long – that spews out through the surrounding galaxy at close to the speed of light, causing havoc for anything that lies in its path.

So much for our once quiet, pleasant and orderly universe.

Quasars are dangerous beasts, and there are a lot of them out there: the latest estimate is around 200,000. Boasting the most massive black hole is the quasar OJ 287, which harbours a hidden monster weighing in at 18 billion times the mass of the Sun. It's being circled by another black hole – with a mass of 'only' 100 million Suns. The two are expected to merge in the incredibly short astronomical timescale of 10,000 years.

Time is of the essence for quasars. They are products of the young universe: baby galaxies on a feeding frenzy of gargantuan proportions. It's no coincidence that we see them far out in space, far away in time, because we are looking at the products of the young cosmos, when it was unstable, fiery and active. But once the supply of food to a quasar's stomach has been gobbled up, all goes quiet. That's why there are no quasars in our galactic neighbourhood today.

All this, however, might change. Instead of having a redshift, our neighbour galaxy – the giant spiral in Andromeda – shows a blueshift. Which means that it is approaching us. And it's very much on the cards that our Milky Way and Andromeda will eventually collide, and merge with each other.

The result will be a new elliptical or disc-shaped galaxy, called Milkomeda.

For individual stars like the Sun, the outcome won't be as dire as it sounds. Stars are so far apart that – even in merging galaxies – they can pass each other safely, like orderly passengers at a crowded airport.

But the two supermassive black holes will feel an irresistible attraction to each other. They'll tango closer and closer, until they eventually merge into one. The new mega-black hole will find a sumptuous feast in the gas and debris pouring in from the collision.

The outcome? A new quasar: right on our doorstep. But no worries. It won't happen for at least four billion years!

Chapter 12

IN THE BEGINNING

The creator of the universe was a most unlikely person. Monseigneur Georges Henri Joseph Edouard Lemaître was a 37-year-old Belgian Roman Catholic priest – and an astronomer. In 1931, he concluded: 'The world has proceeded from the condensed to the diffuse … We can conceive of space beginning with a primaeval atom, and the beginning of space being marked by the beginning of time …'

Now – where had *that* come from?

Flash back some twenty years. Researchers at the Lowell Observatory in Flagstaff, Arizona, had just taken delivery of a new toy: a spectrograph to split up the light from celestial objects to find out what they were made of. Spectra are wonderfully useful to astronomers: not only do they reveal odd facts such as an object's composition; they can also tell how fast it is moving.

Think when you last heard an ambulance racing towards you, siren blaring, and then speed off into the distance. Remember how the siren's pitch was higher as it approached, then dropped as it receded? The sound waves were bunched up on the approach to create a higher frequency – which dropped when the ambulance sped away, and the sound waves trailed behind.

You've just experienced the Doppler Effect – recognised by the Austrian physicist Christian Doppler in 1842. Three years later, the Dutch meteorologist Christoph Buys Ballot confirmed Doppler's findings. On the Utrecht–Amsterdam railway line, he hired a train, occupied by an orchestra, to play a specially calibrated note as they sped past a group of musicians with perfect pitch. A sensational demonstration of the Doppler Effect!

Light experiences the Doppler Effect too. A spectrograph spreads out light into a rainbow of colours: red, at one end, corresponds to low frequencies of light, like the bass notes of a musical instrument; blue and violet light are the treble notes. Various atoms in a celestial body cause dark lines in its spectrum, like the keys on a clarinet. And – as with Buys Ballot's musical note – the spectral lines of a celestial body are shifted if it's moving: to higher, or bluer, frequencies if it's approaching; and to the red if it's receding.

At Flagstaff, the director, Vesto Melvin Slipher, used the technique to check out if the 'spiral nebulae' (which we introduced in Chapter 5) rotated. He believed – like many astronomers of the time – that these were solar systems in the process of formation, and thought they would shed light on the making of planets. In the early twentieth century he had no idea that they were distant galaxies like our Milky Way, made up of billions of stars.

Slipher discovered that the spiral nebulae weren't spinning noticeably. Instead, the spectra showed that almost all of them were rushing away from us at very high speeds, producing Doppler's 'redshift'.

Then Slipher came up with his masterstroke. He made an immense mental leap. Far from being nearby planetary systems in the process of formation, the spiral nebulae were far greater denizens of the cosmos. In 1917, he wrote: 'It has for a long time been suggested that the spiral nebulae are stellar systems seen at great distances. This is the so-called "island universe" theory, which regards our stellar system and the Milky Way as a great spiral nebula which we see from within. This theory, it seems to me, gains favour in the present observations.'

He presented his redshifts to the American Astronomical Society – who gave him a standing ovation. But no one actually knew what to make of them.

It took a giant like Edwin Hubble to come up with an explanation. Hubble had started off life as a brilliant law student – in deference to his father – and eventually a Rhodes Scholar at Oxford. Along the way, he dabbled in athletics and amateur boxing and was a tank-driver in the First World War. But at college, he was drawn to astronomy; and soon his passion for the heavens took over. 'I would rather be a second-rate astronomer than a first-class lawyer,' he once declared. 'All I want is astronomy.'

After being discharged from the war effort, he joined the great Mount Wilson Observatory above Pasadena in California. His arrival, in 1919, coincided with that of the huge Hundred-Inch Telescope: then the biggest in the world.

The brilliant young astronomer was unleashed onto the mighty eye on the sky, and soon pinned down the nature of the spiral nebulae. Hubble proved that they were definitely made up of millions of stars. These were indeed – in the parlance of the time – 'island universes', meaning other Milky Ways. It was a sea-change in the way astronomers thought about the universe: it had suddenly become much, much bigger.

In 1928, Hubble embarked on a new campaign with the Mount Wilson telescope – to measure the distances to what he called the 'extragalactic nebulae' (which, mercifully, became shortened to galaxies).

He was joined in his endeavours by a most unlikely accomplice. Milton Humason – who'd dropped out of school when he was fourteen – began his working life as a mule driver, bringing supplies up the long, rutted track to the top of Mount Wilson.

Humason eventually married the daughter of the observatory's engineer. And he so much loved the place that he managed to secure a menial job as a janitor there. Humason later recalled: 'My own first meeting with Hubble occurred when he was just beginning observations at Mount Wilson. I received a vivid impression of the man that night that has remained with me over the years.'

Humason continued: 'His tall, vigorous figure, pipe in mouth, was clearly outlined against the sky. A brisk wind whipped his military trench coat around his body and occasionally blew sparks from his pipe into the darkness of the dome. The confidence and enthusiasm which he showed on that night were typical of the way he approached all his problems. He was sure of himself – of what he wanted to do, and how to do it.'

As well as being an astronomy enthusiast himself, Humason was also a brilliant observer. Although untutored, he was soon hired as a night assistant, working with Hubble, the master of the galaxies. He applied his meticulous observing skills to measuring their spectra.

Hubble and Humason pushed the boundaries of their mighty telescope, and discovered – like Slipher – that almost every galaxy was speeding away from us. Nearly all of them boasted a redshift.

The team's early measurements of the distances to their galaxies were nothing like as precise as those today. They relied on a number of brilliant interstellar beacons – luminous variable stars that had been calibrated in our own Galaxy, and would serve as 'standard candles' millions of light years away. At least Hubble and Humason could roughly work out a galaxy's distance.

They discovered that the further away a galaxy lies, the greater is its redshift. What was going on? There were many scratched heads at this point. Perhaps light was getting 'tired' – losing energy, and becoming redder – by travelling the colossal distances across space. Or was something more fundamental taking place?

There was indeed.

Having studied physics and astronomy, Georges Lemaître achieved the ambition of his life in September 1923: to be ordained as a Catholic priest. But he never lost his passion for science, keeping his two heavenly interests in separate compartments. His training for the priesthood allowed him time to review the scientific literature. He also studied at Cambridge and in the United States, where talk of redshifted 'nebulae' was all the buzz.

In 1927, Lemaître published a paper catchily entitled: 'A homogeneous universe of constant mass and growing radius accounting for the radial velocity of extragalactic nebulae'. He had predicted that the universe was expanding, fully two years ahead of Hubble and Humason, who came to the same conclusion in

1929. But Lemaître's paper was published in an obscure Belgian journal – whereas the American team went into print in the influential *Proceedings of the National Academy of Sciences*. People – understandably – latched onto Hubble and Humason.

It was now established that the universe was expanding. But why? What had caused it? In 1931, Lemaître made his ultimate prediction. The universe, he maintained, had a precise beginning in a 'primaeval atom'.

Publishing in the prestigious journal *Nature*, he wrote of 'a Cosmic Egg, exploding at the moment of the creation'. Later in life, he would write:

> The evolution of the world can be compared to a display of fireworks that has just ended: some few red wisps, ashes and smoke. Standing on a well-chilled cinder, we see the slow fading of the suns, and we try to recall the vanished brilliance of the origin of the worlds.
>
> The whole universe would be produced by the disintegration of this primaeval atom. It can be shown that the radius of space must increase. Some fragments retain their products of disintegration and form clusters of stars or individual stars of any mass.
>
> It is quite possible that the expansion has already passed the equilibrium radius, and will not be followed by a contraction. In this case ... the suns will become colder, the nebulae will recede, the cinders and smoke of the original fireworks will cool off and disperse ...
>
> ... Whether this is wild imagination or physical hypothesis cannot be said at present, but we may hope that the question will not wait too long to be solved ...

It was an incredibly bold prediction. Was it a result of Lemaître's religious faith, in which unexplained miracles are allowed to happen? In the event, it didn't at first go down well with most physicists, who adhere to the principle of causality. If there was a sudden primaeval atom or cosmic egg, what created it? And what came before the entity that created the cosmic egg … and before that?

Leading the dissidents was maverick astrophysicist Fred Hoyle. They don't make scientists like him any more – regrettably. Today, due to strapped budgets and academic politics, science is a matter of teamwork and consensus. Groups publish their findings collectively. There has to be overall agreement, and seldom does anything sensational or world-shattering emerge. The days of an individual voice in science are numbered.

Not so with Fred (nobody would have dared to call him 'Hoyle'). In the middle of the last century, this blunt Yorkshireman was never afraid to speak his mind. A theoretical physicist, he was wrong as often as he was right. He was also a gifted sci-fi writer, and both of us remember cowering behind our sofas as children when his series *A for Andromeda* was screened on TV.

In 1948, Fred – aided by colleagues Hermann Bondi and Tommy Gold – came up with a completely new theory for the origin of the universe. In their view, the universe had no end, and no beginning. Although expanding, the cosmos stayed in perfect balance, like a washing-up bowl that's kept topped-up by a trickle from a tap. The 'tap', in this case, was the continuous creation of matter from energy. Fred's team called their alternative theory of the origin of the universe 'The Steady State'.

In his theory of special relativity, Albert Einstein had already predicted that matter and energy were interchangeable. In his famous equation $E=mc^2$, energy (E) and mass (m) can be converted into each other. Hoyle and his team's demands on the universe were hardly excessive. To maintain the Steady State, just one new hydrogen atom would need to emerge in a volume the size of a wine bottle – once every billion years.

Fred – so convinced of his Steady State – went on BBC radio in 1950, where he dubbed Lemaître's theory of the exploding primaeval atom the 'Big Bang'. The talk was so popular that a BBC audience panel of 4,000 listeners named Fred broadcaster of the year. As the *Daily Graphic* proudly boasted: 'Pollsters found him more popular than Bertrand Russell, Dr Joad, Tommy Handley or even Wilfred Pickles ... Hoyle's challenging ideas have been helped by his homely accent, his dogmatism, his confidence ...'

The irony was that the beast he wanted to destroy – the idea of a one-off creation of the universe – stuck forever in the minds of the public. Lemaître's 'cosmic egg' had, thanks to outspoken Fred, become the catchier Big Bang theory.

Fast-forward to the early 1960s. Two astrophysicists had just set out on a survey of the peripheries of our Milky Way Galaxy. They were wrestling with a cumbersome antenna – a giant metal horn twenty feet (6m) across, which had been used to receive radio signals from the first communications satellites.

In the flat fields of Holmdel, New Jersey, the duo – Arno Penzias and Robert Wilson – sensed that things weren't going right from Day One. Robert Wilson recalls his bafflement: 'When we turned it on, we knew immediately that there was a signal in there that we didn't understand. Maybe it was junk from New York City ...'

Was it electrical interference from the Big Apple? Or could the explanation lie closer to home?

'There were a pair of pigeons living in the antenna,' remembers Wilson, 'so the inside of the horn was covered in pigeon droppings. Arno and I in our white lab coats got up there with a broom and cleared out the droppings, but nothing seemed to change things.'

The pigeons were caught, released and urged to find a new nest. Unfortunately – being homing pigeons – they were soon back. The team described that this time they were discouraged by 'more decisive means'.

Arno Penzias tells us about the gory solution. 'We found that the most humane way of getting rid of them was to get a shotgun – and from very close range, just killed them instantly. It's not something I'm happy about but it seemed the only way out of our dilemma.'

But despite the demise of the hapless pigeons, the hiss went on. The duo meticulously cleared all the surfaces of their horn yet again – but the signal still wouldn't go away. It was spread out with a fantastic degree of uniformity across the sky, showing no change with day or night, or with time of year.

Penzias despaired: 'We frankly did not know what to do with our result – knowing, at the time, that no astronomical explanation was possible.'

As physicists, Penzias and Wilson were also feeling a tad guilty that their meticulous calibration of the antenna hadn't seemed to work. So they started to write a detailed troubleshooting report on their technical procedures, and put the hiss behind them.

In a casual phone call to a colleague one day, Penzias mentioned the hiss. It rang bells with him – in the shape of a preprint of a research paper from nearby Princeton University. Physicist Robert Dicke and his team were on the trail of evidence that our universe had been born in a Big Bang.

Penzias invited the team over to Holmdel. Dicke looked at the horn antenna's result, turned to his team, and famously chuckled: 'Boys – we've been scooped.' Because, staring at them in the face, was proof of the origin of the universe.

Wilson – a Steady State fan – was gobsmacked.

'When we talked with Dicke, we found out that they were looking at a theory of the Big Bang. In such a situation, the universe would be hot – but if it was hot in the beginning, it would cool down, and the radiation that filled it at the beginning would be visible now as radio waves.'

The mysterious hiss was radiation that had been cooled by the expanding universe to about three degrees above Absolute Zero, the temperature at which atomic motion stops – around minus 273°C. Penzias and Wilson had inadvertently discovered the afterglow of creation itself. And for their painstaking work, they would be awarded the Nobel Prize in 1978.

Not everyone was happy with the result. Fred Hoyle, Hermann Bondi and Tommy Gold still clung on to a 'modified' Steady State theory. And Heather remembers one of her professors at Oxford – Dennis Sciama – recalling the day the Steady State died. 'I was so sad that I cried. Because it was such a beautiful theory.'

Georges Lemaître lived until 1966, the year after Penzias and Wilson discovered their cosmic hiss. He must have died a happy man, knowing by then that his revolutionary theory of the origin of the universe had been vindicated.

Today, the evidence for the Big Bang is overwhelming. Backtrack the motion of the expanding universe, and you come to a point – 13.8 billion years ago – when a tiny speck of brilliant light appeared. Hotter and denser than anything we can imagine, inside this fireball was the whole of space. And with the creation of space came the birth of time, and the great cosmic clock started to tick.

Why? We might never know. Our present physics breaks down when we're up against the point of creation.

And the obvious question: what was there before the Big Bang? Astronomer Royal Sir Martin Rees says: 'We can't really say what happened before the Big Bang. Time – in the sense we understand it – began with the Big Bang, and didn't really exist before. "Before" and "after" may not really make sense in this context because that presupposes the idea that we have clocks that tick away steadily, and that idea might have to be jettisoned.'

Secondly, what caused the Big Bang? Again, there is no satisfactory answer. Physicists currently believe that empty space on its smallest scales – a trillion-trillionth the size of an atom – consists of 'cosmic foam', boiling and bubbling like the tiniest waves and eddies on the ocean. These ephemeral bubbles may spontaneously generate baby universes, most of which live for only a fraction of a second.

'Maybe our Big Bang was not the only one,' reflects Rees. 'What we call our universe is not everything there is but just one part of a grander cosmos in which there are many other universes – maybe an infinite number. We could be all part of a "multiverse", produced by a series of big bangs in some grand eternal structure.'

Although the Big Bang undeniably happened, there are limits to our knowledge. At present, we can only probe backwards to fractions of a second after the event, and not to the Big Bang itself. Admittedly, we can push right back to 10 million-trillion-trillion-trillionths of a second after creation – but that's where the screen goes blank. 'Conditions get beyond the range we can actually simulate here on Earth in our laboratories,' explains Rees. 'When we get back further and further, we get less certain that we understand the nature of the Big Bang.'

One thing is for sure: the infant universe hit the ground running. As soon as it was born, it started expanding – not into anything, but throughout, because the universe is everything and everywhere. However, this was as nothing compared to what was about to happen. Suddenly the universe blew up. In practically no time at all, it grew a hundred trillion trillion trillion trillion times. This phenomenal growth – cosmic inflation – made the original Big Bang seem about as sensational as a hand grenade going off in a nuclear war.

The 1979 theory of inflation was the brainchild of Alan Guth, then a 32-year-old particle physicist working at the Stanford Linear Accelerator Center in California. His goal was to find answers to why the universe is so smooth on its largest scales, and what created the forces that permeate the cosmos today. Guth explains: 'At high energies, there really should exist a very peculiar form of matter that would actually turn gravity on its head – and cause it to become repulsive.'

The seconds-old universe was not only a blazing inferno – it was a battleground. The energy fields unleashed by inflation spontaneously created matter. And the young universe was nothing if not experimental with its very early matter. To get a feel for the kinds of exotic dwellers swimming around in the infant cosmic sea, we can search the innards of particle accelerators on Earth.

But the universe, with its far vaster energy reserves, could go one better. Wimps, leptons, quarks, X-bosons, gluons and gravitons rubbed shoulders with Higgs bosons, neutrinos, primordial black holes, cosmic strings and magnetic monopoles. Many of these early creations rapidly decayed, or changed into other particles. It was an era of total turmoil.

The fledgling universe was expanding at breakneck speed. And as the cosmos grew, it continued to cool down. At the age of one second, its temperature was a 'mere' 10 billion degrees. As the heat continued to ease off, subatomic construction kicked into action.

Gregarious particles called quarks came together to make protons and neutrons. In doing this, the young universe had made its first element – for a proton is the nucleus of the lightest element, hydrogen. By forging hydrogen, the universe had created its most fundamental building block – even today, hydrogen makes up 77 per cent of the cosmos. This is the 'H' in water's H_2O, and it is the most abundant atom in our bodies.

By the end of the first three minutes, the process of element creation was nearly over – the particles were, by then, being driven relentlessly apart by the continuing expansion. There was just time for two protons and two neutrons to come together to create the centre of a helium atom. Helium now constitutes 23 per cent of the universe. In everyday life, it's best known as the gas that fills party balloons – we owe that to the Big Bang!

The quantities of hydrogen and helium, incidentally, provide added evidence that the Big Bang really did take place. Knowing the temperature, density and initial expansion of the universe, physicists can calculate which elements should have formed and in what proportions – and the results match the predictions exactly.

The universe now settled down into a much more sedate phase. It was filled with an opaque fog of subatomic particles, glowing brilliantly as it gradually cooled.

Then, one day, the fog abruptly cleared. A third of a million years after the Big Bang, the universe became dark. 'The temperature of the universe had by then dropped to a few thousand degrees – about that of a tungsten lamp,' explains Martin Rees. At this temperature, the electrons in the cosmic soup slowed down so much that they could be grabbed by the massive protons and neutrons. They combined to make the first chemical elements of the cosmos. And radiation was at last allowed a clear passage.

Astronomers are fascinated by the light that streamed from the primordial fog. Now stretched out to radio wavelengths, it's the radiation that Penzias and Wilson detected back in 1965. And researchers have launched several satellites to probe this relic of the early universe in ever more detail. 'It's a direct fossil of the early universe,' observes Martin Rees.

The latest mission – Europe's *Planck* spacecraft – was launched in 2009, looking for clues as to what drove the early universe. Essentially the ultimate space thermometer, *Planck* held the record for being the coldest object in space, until its coolant of liquid helium ran out in 2013.

In March 2013, the *Planck* team published their findings, with a map of the universe when it was just 370,000 years old. This ultimate 'baby picture' of the cosmos reveals subtle fluctuations in temperature, that show the gas in the early universe 'curdling' into the clusters of galaxies that populate our universe today.

But – after this point – our view of our early universe becomes blurred. Dealing with the ferocious conditions just fractions of a second after the Big Bang itself is, ironically, more of an exact science – physicists can at least compare them to what happens in a particle accelerator. Martin Rees explains: 'At this point, the universe became a dark place, and remained dark until the first stars and the first galaxies formed and lit it up again. One of the key questions for cosmology now is how long the Dark Age lasted.'

The amazingly far-sighted Hubble Space Telescope has looked deep into space – and time – to view the first galaxies, born by the time the universe was roughly a billion years old. These pictures reveal that baby galaxies look nothing like their suave, sleek counterparts of today. They're scruffy, ragged and garish, packed with searingly hot blue stars glaring uncompromisingly at their companions.

Other incredibly distant members of the early cosmos were the quasars (which we encountered in the previous chapter): galaxies containing supermassive black holes, whose gravity triggers explosive and catastrophic outbursts.

Make no mistake: even after the Big Bang, the universe was still a very violent place.

As the universe matured, the pressure eased off. The first frenetic epoch of star formation slowed to a respectable pace. Stars – originally made from only hydrogen and helium – processed new elements in their central nuclear reactors, which led to greater possibilities for the cosmos. Forging ever more complex elements in their fiery cores, stars produced the atoms – like carbon – that were destined to create life.

But it might never have happened. 'Our universe does, in a sense, seem to be rather special,' observes Martin Rees. 'We could readily imagine other universes which are – as it were – sterile. They may not contain stars, or they may not live long enough to allow for complex evolution or for life to develop.'

The Big Bang led to the birth of galaxies, stars, planets – and to the evolution of a lifeform on planet Number Three around an anonymous star in an average galaxy. A species that has come to grapple with the nature of its ultimate origins, almost 14 billion years ago.

But – according to Rees – it's early days yet. 'One thing that cosmology teaches us is that our emergence depended on physical processes which we can trace right back to the Big Bang. Another thing it teaches us is that we are still near the beginning of cosmic evolution.'

He adds: 'Most of the cosmic course is still to run, and I think we should see ourselves not in any sense a culmination of the process, but in a position to speculate about what may happen over the billions of years that lie ahead.'

The future beckons ...

Chapter 13

THE DARK SIDE

Our universe is controlled by the powers of darkness. Two mighty invisible forces are locked in combat to control the planets, stars and galaxies. The outcome will determine the ultimate fate of the cosmos ...

That paragraph may sound like advertising hype for the latest computer game. But – incredibly – it's true. Astronomers now have cast-iron evidence that all the majestic galaxies we see in the universe, with their brilliant stars and gaudy gas clouds, together account for only a few per cent of the 'stuff' that makes up our cosmos.

Over nine-tenths of our universe is missing: presumed 'dark'. And it's the dark material that's in control.

Astronomers didn't come to such an astounding conclusion easily. The evidence has actually been around for eighty years; but only with the latest mighty telescopes have they been forced to admit the overwhelming power of the unseen cosmos.

When they survey the geography of the universe on the biggest scales – a panorama stretching out over billions of light years – astronomers have found that the visible galaxies are like the lights of Alpine villages, seen as we fly over them by night. From our aircraft, we can't easily work out why these communities lie where they do; or why their roads link together in particular ways. We need to know about the dark mountains between the villages before we can understand the geography of the lights.

It's the same with the universe. Only by exploring the dark forces can we understand the shape and size of the galaxies, and the clusters – 'cosmic villages' – that they inhabit.

Welcome to the dark side …

The father of the dark universe was a maverick Swiss astronomer named Fritz Zwicky. He grew up in Zurich, a great melting pot of a city that he shared with the psychoanalyst Carl Jung, the Irish novelist James Joyce and the founders of the subversive Dada art movement. Albert Einstein taught at the college where Zwicky studied; while Zwicky's next-door neighbour was Lenin.

In the 1920s, Zwicky moved to California, where he was swept away by the excitement surrounding the new Hundred-Inch Telescope. Edwin Hubble and his team were laying bare the secrets of the distant universe, and it provoked Zwicky's fertile mind to wild new ideas beyond the ken of his more conservative American colleagues.

It wasn't just his unorthodox imagination, though, that kept him an outsider. Unlike Einstein, who moved to the United States a few years later, Zwicky refused to become an American citizen. He never bothered with becoming fluent in English. A standing joke was that 'Fritz speaks seven languages – all badly'.

And he didn't give up on his somewhat direct tongue. Zwicky never suffered fools gladly – and he regarded most of his colleagues as fools. He called them 'spherical bastards'. Why? 'Because they are bastards, when I look at them from any direction.' On one occasion when he had invited some students to dinner, Zwicky's wife – inured to her husband's vocabulary – opened the door and called out, 'Fritz, the bastards are here!'

Away from his workmates, Fritz Zwicky showed a very different side to his personality. After the Second World War, he took a personal hand in re-homing orphans whose parents had died in the conflict. Star-expert Cecilia Payne-Gaposchkin remembered him fondly. Herself an outsider – as a female scientist, an immigrant from Britain and the wife of a Russian – Payne-Gaposchkin described Zwicky as: 'the kindest of all men, with a deep concern for humanity'.

Like Fred Hoyle – whom we met in the previous chapter – Zwicky refused to toe the scientific party line. In the eyes of Payne-Gaposchkin, he was: 'the last of the scientific individualists, a breed that is dying out in the age of teamwork'. Though Zwicky's name has unjustly remained in the margins of astronomical history, he was a towering genius when it came to breakthrough ideas.

Not all of them worked out. Zwicky was observing one night with the Hundred-Inch Telescope at Mount Wilson, when he found that the sky was heaving with atmospheric turbulence. He instructed his night assistant to grab a gun and shoot into the churning air, hoping that the blast would calm down the turbulence. It didn't work – but the incident demonstrated Zwicky's tangential approach to problem-solving.

Later in life, he may have been more successful – in being the first person to launch an object into interplanetary space. In 1957, Zwicky packed three aluminium balls on top of explosive charges, and launched them to a height of 87 kilometres on a small rocket. The explosives detonated. One of the balls headed back into the atmosphere and burned up; but the others were never seen again – presumably shot up into space. The balls were travelling fast enough to break free of Earth's gravity, and become the first human-made objects to go into orbit around the Sun.

But it was high-speed galaxies – rather than rocketing balls – that would put Zwicky on the trail of the dark universe.

Hubble had just identified distant fuzzy blobs as giant 'star-cities', like our Milky Way. But how heavy were these galaxies? Most astronomers took a simple view, assuming galaxies were just made of stars like the Sun. If you measure a galaxy's brightness as – for example – a billion times more luminous than the Sun, then the galaxy would be a billion times more massive than the Sun, too.

But Zwicky was always thinking outside the box. Suppose a galaxy contained 'stuff' that wasn't in the form of stars? He concluded: 'Estimates of the masses of galaxies from their observed luminosities are therefore incomplete and can at best furnish only the lower limits for the values of these masses.' (Zwicky, by the way, referred to galaxies as 'nebulae' – we've inserted the modern name to make his quotations clearer.)

He cast around for a way to measure a galaxy's weight directly. The best 'weighing machine' in the cosmos is gravity: we can work out the mass of a planet, for instance, by watching a moon orbiting under the influence of its gravity.

When it comes to the cosmic scale, Zwicky didn't observe galaxies actually orbiting each other. Instead, he looked at galaxy clusters – swarms of galaxies, held together by the invisible threads of gravity. Our own Milky Way lives in a petite cluster called the Local Group – with a few dozen galaxies swimming around one another. But other galaxy clusters are vast shoals, containing thousands of the beasts.

In 1933, Zwicky turned his attention to the Coma Cluster, a giant cosmic conurbation of more than a thousand galaxies. By breaking up the light from the galaxies into a spectrum, Zwicky could tell how fast they were moving.

He was astonished. The galaxies were haring through space far more speedily than anyone had expected; rapidly enough to break free from the gravity of their neighbours. The Coma Cluster should long since have scattered across the universe.

Zwicky took stock. There was only one answer. To hold the Coma Cluster together, the gravity of the stars in its galaxies must be supplemented by the gravity of something else – something unseen – which was far more powerful. Zwicky named it 'dunkle Materie'; in English, 'dark matter'.

But nobody listened to the maverick Zwicky. He was to be a voice in the wilderness for half a century … until Vera Rubin stumbled across another cosmic enigma.

Rubin also had problems with her colleagues, first as a woman who'd graduated from an all-female college in 1948, hoping to do research at Princeton University – only to find that women were banned from its graduate school. And it didn't end there. Her first conference was a disaster. Accompanied by her one-month-old son, Rubin travelled to a meeting of the American Astronomical Society in Philadelphia in 1950 to present a paper on the nature of the universe.

'I got up, gave my paper,' she later recalled, 'with the enthusiasm of youth. My paper was followed by a rather acrimonious discussion.' More forthright friends saw her treatment as humiliating.

But Rubin was determined; like Zwicky, she thought unconventionally. 'I did not understand how far off of the establishment I was going. It never occurred to me that this was not the kind of thing that someone else could have done.'

Like many female astronomers of the 1950s and 60s, Vera Rubin had to trail around universities, following her physicist husband's postings. Eventually, she ended up at Washington's Carnegie Institution, in the strangely named Department of Terrestrial Magnetism – strange, because she had elected to study galaxies.

Here, Rubin formed a dream-team with Kent Ford. He had built a highly sensitive spectrometer, which could prise apart the light from even the faintest regions of dim galaxies. And so she could fulfil her dream of unravelling the secrets of remote depths of the universe.

'Observing is spectacularly lovely ... I enjoy analysing the observations, trying to see what you have, trying to understand what you're learning. It's a challenge, but a great deal of fun. It's not only fun, a lot of it is just plain curiosity – this incredible hope that somehow we can learn how the universe works. What keeps me going is this hope and curiosity.'

Rubin focused her attention on spiral galaxies – great cosmic pinwheels like our Milky Way. With Ford's instrument, she could measure how fast they were spinning. And – like Zwicky – she came up against an apparent violation of a cosmic speed limit.

Astronomers had thought that most of a spiral galaxy's mass was in the centre – where the stars are most densely packed. In that case, the outer regions should behave a bit like the planets of the Solar System, which are subject to the gravity of the central Sun: Mercury, the innermost planet, rushes around fastest, while remote Neptune merely creeps along its orbit.

But Rubin found that the stars in the outermost arms of the spiral galaxies kept up a similar speed to their cousins close to the galaxy's core. Something had to be holding in the outermost stars, which – otherwise – were speeding so fast that they should escape the galaxy altogether. 'What you see in a spiral galaxy,' Rubin concluded, 'is not what you get.'

Was there something wrong with the theory of gravity? Or was there another explanation? Rubin believed the latter – the stars in the galaxy were reined in by the overriding gravity of some kind of invisible matter.

Rubin had little support at first. But then a team of Estonian astronomers came to a similar conclusion. They were investigating not just the motion of stars within galaxies, but the way that small galaxies orbit around bigger ones. Every galaxy they studied had to contain a lot more matter than just the stars and gas we can see. In 1974, their leader, Jaan Einasto, announced at a winter school in the Caucasus: 'All giant galaxies have massive coronas. Therefore dark matter must be the dominating component in the whole universe – at least 90 per cent of all matter.'

Rubin and Einasto were also saying that this dark matter wasn't concentrated in a galaxy's centre – like the stars – but forms a great sea of unknown material that permeates the whole galaxy, and out into intergalactic space beyond.

The clincher came in the 1980s, when astronomers found dark matter performing its most amazing trick yet – acting like a vast telescope lens deep in space.

Einstein had calculated that gravity has the power to bend a beam of light. And it was the hero of this chapter, Fritz Zwicky, who – back in 1937 – said that a galaxy could pull on the light passing by, focusing it like a glass lens. The heavier the galaxy, the more powerful the lens. With amazing foresight, he wrote: 'observations on the deflection of light around galaxies may provide the most direct determination of galactic masses'.

Fifty years later – and a decade after his death – Zwicky was proved right. In 1987, a team of astronomers was probing the sky with a giant telescope on the peak of Mauna Kea, Hawaii, when they stumbled over a very strange sight. Surrounding some nearby galaxies, they found peculiar banana-shaped arcs. Each arc has turned out to be the distorted image of a distant galaxy, focused by the gravity of the foreground star-city. And the eagle eye of the Hubble Space Telescope has now turned up gravitational lenses by the hundred.

Vera Rubin once said, 'We are still groping, as if we are in a black room trying to make a black puzzle.' Now, gravitational lenses are illuminating our perspective of the dark universe. They are the best weighing machine yet for distant galaxies.

And these invisible lenses have proved that dark matter is the ultimate architect of the cosmos. Its gravity – astronomers are now finding – links galaxy clusters together in long strands, separated by vast empty regions of space.

'There's a filamentary pattern,' explains cosmologist Lawrence Krauss, 'a cosmic web containing galaxies, and clusters of galaxies, that light up the universe. In fact, on the larger scales the universe kind of looks like a sponge.'

For years, Krauss has been in the vanguard of scientists trying to grapple with the dark universe. So, what is the mysterious substance that is shaping the cosmos?

'Dark matter is weird,' Krauss concedes, 'because we don't understand it at all. It's clearly not made of the same stuff you and I are made of.'

And it's not just 'out there'. Dark matter fills galaxies; it sloshes around inside the Milky Way; it's here, in the Solar System. You are surrounded by dark matter right now; and it's permeating your body.

'You can't push against it, you can't feel it,' continues Krauss. 'It's a ghost-like material that'll pass right through you as if you didn't exist at all.'

The best guess is that dark matter consists of some kind of tiny particle – like the protons and electrons that make up atoms. Scientists have even devised a name for these unsociable minnows: 'weakly interacting massive particles'.

The 'weakly interacting' part is clear enough – they're shy of getting involved with ordinary matter. To scientists, they are 'massive' because they're far heavier than an electron – maybe as massive as an atom of iron – though that's minuscule, of course, in everyday life.

'Weakly interacting massive particle' is a bit of a mouthful to trot out in conversation. So, although we're not keen on acronyms, in this case we'll make an exception: it's just so wonderful to say that 'WIMPs' are in charge of mighty galaxies!

And now the great WIMP hunt is on. To paraphrase the great spy story *The Scarlet Pimpernel*, 'They seek it here, they seek it there, those scientists seek it everywhere ...'

Some astronomers are looking far out into space to pick out the remains of WIMPs that disintegrate in the core of galaxies: there are tantalising hints in the form of unusual radiation coming from the centre of our own Milky Way.

'But you have to realise it's not just "out there",' says Krauss. 'It's here – it's in this room – and it's going through you and me. So we can build detectors deep underground to look for the individual particles.'

So the WIMP hunt also takes us to Boulby Mine in North Yorkshire. The mine was dug to extract potash – a valuable fertiliser – from more than a kilometre underground; so deep that the temperature here is a sweaty 40°C. But scientists hope to reap an equally rich harvest from a futuristic lab that's been built in the mine's lowest levels.

The experiments down the pit are designed to detect the very rare occasions when a WIMP actually hits an ordinary atom. The thick layers of rock shield their experiments from cosmic radiation. To cut out interference from radioactivity in the surrounding rocks, the experiment is encased in sheets of inert old lead, from the roof of Salisbury Cathedral!

Even more audaciously, physicists at the giant CERN particle accelerator near Geneva are trying to create WIMPs from scratch. History was made here in 2013, when scientists discovered a long-predicted massive particle – the Higgs boson – in the debris spewing from the collision of two high-speed beams of protons.

The next big breakthrough here could be the successful ensnarement of the WIMP. The collider at CERN re-creates – in exquisite miniature – the fireball of the Big Bang, where the dark matter thronging our universe was made.

Astronomers calculate that the WIMPs spawned in the Big Bang outweigh the ordinary matter in the universe five times over; and the gravity of these WIMPS controls the geography of the cosmos.

When it comes to the *history* of the universe, though, it's a different story. Astronomers have found that dark matter is upstaged by a rival – and even more mysterious – power: the force of dark energy ...

The trail again leads back to our old friend, the unsung genius Fritz Zwicky. In 1937, he was taking pictures of an obscure galaxy called IC 4182, in the constellation of Canes Venatici – the Hunting Dogs – when he tracked down an exploding star. And it was no ordinary eruption. The star flared up until it was far more brilliant than the entire galaxy where it resided.

Zwicky coined the term 'supernova' for such brilliant exploding stars. Years ahead of his time, he thought a supernova – of a kind he called 'Type II' – was an old star that was committing suicide, blowing itself apart when it ran out of the nuclear fuel that kept it shining. For many supernovae, that's exactly true (as we saw in Chapter 10).

But the supernova he saw in 1937 was different. It was brighter, and it faded in a very steady fashion. Its light also showed no sign of the great melting pot of new elements that are forged in the death-throes of an old star.

Zwicky named this odd explosion a 'Type I' supernova (astronomers now use the term 'Type Ia'), but even his fertile mind couldn't come up with an explanation. Today, we know he was seeing the death of an old shrunken star called a white dwarf (which we introduced in Chapter 9, in the shape of Sirius's canine companion The Pup).

A white dwarf is made of collapsed atoms, which constitute the most explosive substance in the universe. The 'blue touchpaper' is ignited when a companion star feeds the white dwarf an excessive amount of extra gas. As the stellar powder-keg is pushed over its natural weight limit, the white dwarf destroys itself in an almighty nuclear explosion: a Type Ia supernova.

Zwicky's colleague Walter Baade took the crucial next step. Baade was an immigrant German astronomer, with whom Zwicky – characteristically – had a love-hate relationship. Zwicky once unfairly accused Baade of being a Nazi, and Baade said afterwards, 'I was afraid he was going to kill me.'

Baade worked out the brightness of the supernovae they'd been observing. It turned out that all the Type Ia supernovae reached the same maximum brilliance – today's measurements put that at an astounding five billion times brighter than the Sun.

Bingo! Here was an accurate way to plumb the depths of space. Far-off supernovae are dimmed by their distance; simply by measuring how faint a Type Ia supernova looks in the telescope, you can work out how far away it lies.

For the most distant reaches of the cosmos, the supernova 'standard candle' provides the most accurate way to measure distances. For fifty years, this tape-measure to the galaxies performed brilliantly. With bigger telescopes, and more sensitive electronic cameras, astronomers pinned down galaxies literally billions of light years away.

But, in the late 1990s, things began to go wrong.

As well as measuring galaxy distances, astronomers were clocking their speeds as the universe expands. According to Hubble, a galaxy's speed should be closely linked to its distance (as we explored in the previous chapter). Astronomers had assumed that the universe was just coasting along from the Big Bang. Or, perhaps, it was slowing down a bit as the gravity of galaxies and dark matter put brakes on the expansion.

'By looking at the light from very distant supernovae,' says Bob Kirshner, 'we could tell whether the expansion's been going on at a constant rate, or slowing down as it would be by gravity.'

An expert on supernovae, Kirshner was a key member of a team that was using large telescopes in Chile and Hawaii to check on the continuing growth of the cosmos. 'We thought we were going to measure the slowing-down of the universe,' he explains. 'What that would mean is that distant supernovae would appear just a little brighter than you'd expect in a universe that was just coasting.'

Kirshner's data was being analysed by his colleague Adam Riess. And he was in for a shock when he heard back from Riess.

'I remember it very clearly,' Kirshner recalls. 'I got a phone call from Adam up at Berkeley and he said he's been reducing the data – and it's not lying on the brighter side of the line; it's on the dimmer side. This meant the universe was accelerating. And I found that very upsetting.'

Kirshner had little time to be upset, though. The result was quickly confirmed by a completely different team of astronomers. In one of the greatest sensations in astronomy in our lifetimes, it turns out that galaxies aren't just coasting away from each other; they are speeding apart faster and faster.

Astronomers had discovered the second power of darkness that controls the universe: they call it 'dark energy'.

If dark *matter* had been a joker in the cosmic pack, it was somewhat familiar: dark matter has gravity, and pulls things together. Not so dark *energy*.

'Dark energy,' explains Lawrence Krauss, 'is doing just the opposite – and it's far more mysterious.' Dark energy is a total monster, with some kind of 'anti-gravity' that forces everything in the universe apart from one another.

One scientist, though, had predicted just such a force, over eighty years earlier – but had thrown the idea away. His name? Albert Einstein.

When Einstein applied his new theory of gravity – the general theory of relativity – to the cosmos in 1917, no one knew that the universe was expanding. Einstein was worried that the gravity of all the stars would make the cosmos collapse – so he was relieved when his equations suggested there could be a balancing force of 'anti-gravity'.

But when Hubble discovered the expansion of the universe a few years later, Einstein changed his mind. Says Kirshner: 'Einstein had this idea and later discarded it. He said it was his greatest blunder. But Einstein's waste basket had better things in it than a lot of people are able to generate on their very best days.'

Now, it seems that Einstein was right all along. And, from their latest measurements, cosmologists know that this second dark force is growing in strength. One day, the cosmic repulsion will overcome gravity altogether.

'It means that Einstein's "greatest blunder" was in fact his greatest success,' enthuses Krauss, 'because in the end that's all there will be. It will be responsible for everything continuing to fly apart forever.'

The brilliant universe as we know it today will eventually change beyond recognition – to a mere dark shadow of its former self.

'Dark energy's going to kill the galaxies off,' Krauss elaborates. 'It will do that by causing all the galaxies to recede further and further away from us until they are invisible. The rest of the universe will literally disappear before our very eyes. Not today; not tomorrow; but in perhaps a trillion years the rest of the universe will have disappeared.'

Other scientists are even more pessimistic. According to the Big Rip theory, the forces of dark energy are contagious, and will spread to smaller and smaller scales. After the galaxies are torn away from each other, dark energy will begin to pull the Milky Way apart. It will rip planets away from their sun's gravity; then shred planets and stars themselves.

In the end, even atoms will be broken up by the power of the universe's all-conquering force of dark energy.

But that's a long way into the uncertain future.

'We should be amazed to live now,' says Krauss, 'here at a random time in the history of the universe, on a random planet, at the outskirts of a random galaxy – where we can ask questions and understand things from the beginning of the universe to the end.

'We should celebrate our brief moment in the Sun.'

Chapter 14

A COMET'S TALE

'Of all the meteors in the sky, there's none like Comet Halley.
We see it with the naked eye, and per-iod-i-cally.'

Even Herbert Hall Turner – Professor of Astronomy at Oxford, who penned these lines – got comets and meteors mixed up, as most people do. Meteors are fleeting firecrackers in our skies that flash past in seconds: comets appear to hang like malevolent daggers over our planet.

Hall Turner wrote his ditty a year after Halley's Comet flew by the Earth in 1910; as ever, a once-in-a-lifetime experience that either exhilarated – or terrified – its watchers. In New York, people were so anxious that the comet's tail contained cyanide they bricked up their windows to avoid being poisoned.

Comets have never had a good press. They've always been associated with doom and destruction. Some people have been so traumatised by the sight of these unpredictable celestial visitors that they have hidden behind closed doors when a comet has put in an appearance.

And not without reason. Hammer-blows from space can decimate life on Earth. An asteroid – a comet's first cousin – delivered the death knell to the dinosaurs when it hit the Earth 66 million years ago (as we explain in Chapter 15). And there have been close shaves recently, such as the comet that detonated over Siberia in 1908 – flattening forests for many miles, and killing thousands of reindeer.

But comets have also had their plaudits. History associates them with the birth and death of princes and potentates, as Shakespeare regales us with in *Julius Caesar*: 'When beggars die, there are no comets seen; the heavens themselves blaze forth the death of princes.'

The man who turned comets from superstition into science was the son of a seventeenth-century London soap manufacturer.

Edmond Halley's dad realised that there was a good profit to be made from soap. David Hughes, emeritus Professor of Physics at Sheffield, and an expert on comets, notes the perspicacity of Halley Senior – which was undoubtedly passed on to his son: 'We're talking about the 1660s, when there was the Great Plague of London. And people suddenly realised that stinking and not washing wasn't a good scheme.'

Young Edmond Halley was the complete opposite of the greatest scientist of the time, the introverted Isaac Newton whom we met in Chapter 4 – though the two became good friends. Hughes – a great admirer of Halley – observes: 'He always comes across as a very gregarious, happy, hard-working scientist. He was very ambitious, but a very nice chap as well.'

Halley went on to study at Queen's College, Oxford, funded by his father who gave him the then-enormous sum of £300 a year. Cleanliness clearly pays! Unlike Newton, Edmond Halley didn't want to be closeted up in a college for life. And his father was happy to fund a scheme that Edmond put forward.

Halley realised that three talented astronomers were already measuring up the sky that's visible from Europe – John Flamsteed, at the newly founded Royal Observatory at Greenwich; Gian Domenico Cassini in Paris; and Johannes Hevelius in Danzig. So he proposed sailing south of the Equator and charting the little-known skies down there. Otherwise, he feared he would be merely 'a duck quacking among matchless swans'.

Edmond was also politically astute. On his new chart of the southern sky, he designated a group of stars as Robur Corolinum – Charles's Oak – after the tree in which the king had hidden after the English Civil War. Charles II was so pleased that he ordered Oxford University to grant Halley an MA degree – without Edmond having to do any more work.

Halley stayed in close touch with the king and the government. David Hughes reckons that they earmarked him for reconnaissance jobs abroad: 'Halley went on diplomatic missions to the Mediterranean cities "to look at the fortifications". Reading between the lines, he was admitted as a scientist – but he was also asked by the British government to spy on those places to see what they were up to.'

Edmond Halley's diplomatic skills were to the fore when the Russian Tsar visited London to learn about the latest advances in Europe. After one drunken, uproarious dinner at the diarist John Evelyn's house in Deptford, Halley pushed Peter the Great through a hedge in a wheelbarrow.

It's recorded that Evelyn – who'd been away at the time – was not amused.

And Halley was also a diplomat in the scientific arena. He persuaded the reclusive Isaac Newton to go public with his magnum opus, the *Principia* – which set out the fundamentals of how celestial bodies move under the influence of gravity. In the end, Halley paid for its publication.

The *Principia* finally solved the problem of comets. According to Newton: 'This discussion of comets is the most difficult in the whole book.'

Until then, many astronomers thought that comets obeyed rules of their own. Newton proved that a comet in 1680 was controlled by the Sun's gravity – like the planets – and that it moved in a very elongated orbit which took it out beyond the path of the most distant known planet, Saturn.

David Hughes is amused at Newton's attitude to these celestial eccentrics.

'After he'd calculated the orbit of one comet, dear old Isaac said, "I'm blowed if I'm going to slog my way through a whole gang of others." So he turned to Halley and said: "Look, Ed. I'll give you the observations. Just you go away and calculate these orbits for yourself.'

Newton had collected old observations of twenty-four comets, dating back to 1337. Halley started to analyse them. It wasn't an easy task – because in those days, it took six weeks to calculate the orbit of a single comet.

Hughes, who has spent a lifetime studying comets, empathises with Halley: 'And then, as he was doing this calculation, he realised that three of the orbits he's calculated were very similar. You can imagine him sitting there and saying: I thought I'd calculated twenty-four orbits, but I've calculated twenty-two, because one comet's cropped up three times.

'Whoops! That's periodic, then. It comes back every seventy-six years.'

Halley had seen this comet on his honeymoon – in Islington – during 1682. The other appearances on his list were in 1531 and 1607 – when it was observed by Johannes Kepler in Prague and the Cornish astronomer Sir William Lower. Halley now stretched Newton's new theory, by trying to predict exactly when it would next appear. The answer came out to be 1758.

If he were to be proved correct, Halley wrote: 'Candid posterity will not refuse to acknowledge that this was first discovered by an Englishman.'

The comet did appear as predicted, and Halley's fame was assured.

The celestial visitor we now know as Halley's Comet does indeed put in an appearance every seventy-six years – the last occasion being in 1986, when the European space probe *Giotto* took a close-up look at the cosmic iceberg that forms the comet's core.

Edmond Halley's other activities were equally innovative. He looked into the statistics of deaths, devising the maths behind life assurance. And he went back to his beloved ocean again – inventing the first diving bell, in order to raise cannon from shipwrecks. He even commanded a ship which looked into the performance of the magnetic compass. It was reputed that he could swear like a sea-captain!

Comet research, in fact, formed only a tiny part of Halley's interests – but it's turned out to be his greatest legacy, as these celestial visitors are fundamental to our existence.

Comets are leftover debris from the construction of our Solar System. That's why it is incredibly important to study them: they speak to us of our own origins. In their virgin state, these dirty snowballs – small balls of rock and ice – live in two zones: the Kuiper Belt, close to the orbit of Pluto; and the vast, spherical Oort Cloud stretching a quarter of a way to the nearest star.

And the Oort Cloud may be more than just a swarm of comets. Alan Stern, the scientist leading the *New Horizons* mission to Pluto (Chapter 7), points out that the outer Solar System has been – to put it mildly – beaten up a bit.

'There's evidence for a significantly more massive world beyond Pluto,' he tells us. 'Pluto and its moon Charon were almost certainly formed by a giant impact. Uranus was knocked on its side by a collision, by an object two to five times the Earth's mass.'

Stern believes that the Oort Cloud is thronged with planets that are as massive as the Earth. In the past, similar worlds heading inwards have seriously stirred up the Solar System. And, given his experience of running the *New Horizons* mission, Stern is keen to send a space probe out to find them.

'It would have to go at ten times the speed of *New Horizons*,' he says. 'But with gravity-assist at Jupiter to accelerate it, we could be at the Oort Cloud in thirty years. We have the technology to go there.'

So vast is the Oort Cloud, that it is vulnerable to the gravity of passing stars. This cosmic tweak can dislodge a comet from its perch, sending it speeding into the inner Solar System. David Hughes describes the ride: 'Imagine if I was sitting on one of these cometary nuclei,' he enthuses. 'Then, as it moves in towards the Sun, I'd be moving faster and faster. And the snow inside is getting heated up by the Sun's energy. There's one rather exciting transition, just as I get closer to the Sun than Jupiter: the snow is warm enough to convert to gas.

'And so the surface of the comet suddenly starts emitting gas. When I reach the closest point to the Sun, the surface is disappearing as the snow melts before my eyes.'

By now, the comet is fully fledged. The gas – mainly steam – billows out from the nucleus in a huge halo around the active comet, called the 'coma'. It now sports two tails: one of gas, the other of rocky particles of dust. The gas tail is blue, pushed sharply back into space by the magnetic power of the solar wind. The dust tail, which doesn't contain magnetised particles, can afford to be more leisurely: it curves away from the comet in a yellowish appendage reminiscent of the hindquarters of a ginger cat. Both tails are millions of kilometres long.

The sight of a mature comet in the sky is awesome. Few of us will forget the view of Comet Hale-Bopp in 1997, when it was the brightest object in the sky after Sirius.

What a delightful name for a comet! These cosmic itinerants commemorate their discoverers: a sure way of gaining celestial immortality. The 'Hale' and 'Bopp' in this case are both Americans. Alan Hale is a professional astronomer, whose work involves researching distant stars; but he looks at comets for fun.

'On the night of 22/23 July 1995,' he recalls, 'I had planned to observe two comets. I finished with the first one just before midnight, and had to wait about an hour and a half before the second one rose.'

To while away the time, Hale turned his telescope towards Sagittarius – a constellation rich in star clusters, swarms of tightly packed stars. One of Hale's favourites is the cluster M70.

'When I turned to M70, I saw a fuzzy object in the same field – and almost immediately suspected a comet.'

It was indeed a comet. And Hale contemplates the irony. 'I've spent over four hundred hours of my life looking for comets. And I haven't found anything. Now – suddenly – when I'm not looking for one, I get one dumped on my lap!'

Meanwhile, amateur astronomer Tom Bopp – a shift supervisor at a construction materials company in Phoenix, Arizona – had headed out to the desert with fellow enthusiasts and their telescopes to gaze at the dark skies.

'I was watching M70 slowly drift across the field of view,' he remembers, 'when a slight glow appeared at the eastern edge.'

The group checked their starcharts, but nothing was marked – no galaxy, no nebula, no star cluster. What else would look fuzzy through Bopp's telescope? The group knew only too well. 'We might have something,' they agreed.

Back home, Bopp had a somewhat daunting communication to make: to report a suspected comet, he had to contact the International Astronomical Union. Bopp recalls his reaction. 'When they telephoned and said, "Congratulations, Tom, I believe you discovered a new comet," that was one of the most exciting moments of my life.'

Hale-Bopp was a phenomenon – especially for those living north of the equator. But it was surpassed during the early months of 2007, when Comet McNaught hove into the skies of the southern hemisphere.

The comet, named for Scottish astronomer Rob McNaught at Australia's Siding Spring Observatory, staggered astronomers – and everybody else. McNaught has discovered over fifty comets, but this time he exceeded his own phenomenal record.

Known as 'The Great Comet of 2007', it shone brighter than the brilliant planet Venus. The comet broke all records. It was the biggest comet ever seen: its dust tail stretched an astonishing 150 million kilometres across the Solar System – the distance of the Earth from the Sun.

And what a tail! The huge curving fan split up into dozens of ghostly filaments as the comet ejected bursts of dust. Astronomers described it as 'like seeing eight to ten comets with tails all at once'.

Magnificent though Comet McNaught undoubtedly was, the world still has its soft spot for Halley's Comet. It didn't put on a good display on its last return in 1985–86, because it swung past the Sun a long distance away from the Earth. You could hang on until 2061, when our celestial visitor will be marginally brighter. Best, however, to wait until 2134, when Halley – less than 13 million kilometres away from Earth – will put on a dazzling display in our skies.

People may have felt let down by Halley, and deprived of their 'once in a lifetime' experience. But none of this deterred the European Space Agency (ESA). ESA's *Giotto* space probe – built in Halley's homeland of Britain – pulled out all the stops. The Japanese and Russian space agencies sent missions to the comet, but their results were inconclusive. David Hughes waxes lyrical about *Giotto*: 'This mission was named after the famous Italian artist who painted Halley's Comet in 1301 as the Star of Bethlehem. It passed Halley's Comet at a velocity of 150,000 miles an hour!'

In the early morning of 14 March 1986, *Giotto* beamed back its uniquely intimate close-up of Halley's Comet. The probe flew just 600 kilometres from the comet's nucleus, well inside the coma of gases boiling off Halley.

On TV screens around the world, viewers looked in bafflement at the real-time views, which were in lurid false colour. One of us (Nigel) was faced with the daunting task of commenting on these pictures live from Mission Control for the BBC World Service – fortunately on radio rather than television. It wasn't until the evening that the images were decoded: and *Giotto*'s view turned out to be sharp, clear – and thrilling.

'It was a wonderful picture,' Hughes raves, 'showing great jets of gas activity. We could work out its shape; we could see hills, valleys and depressions on this avocado-pear-shaped cometary nucleus. We could try and work out the way it was spinning; we could work out how much mass it was losing.'

Most of all, scientists were amazed by the colour of Halley's nucleus. It was jet black: as black as coal; as dark as sable velvet. In this 'dirty snowball', the snow is all on the inside, while the dirt coats the outside. Halley's Comet turned out to be a giant cosmic choc-ice.

Brave little *Giotto* didn't fare well in the encounter with its mighty target. Blasted by the dust in the coma, many of its instruments were damaged or destroyed – including the camera. But, undeterred, ESA decided to send their wobbly spacecraft on to another comet. In 1992, it was aimed at Comet Grigg-Skjellerup: a very faint comet that could hardly be seen.

Heather was invited to present a TV programme for the BBC and ESA about the encounter. It struck us that we were making a documentary about a blind space probe intercepting an invisible comet! But – in the event – *Giotto* did detect particles of dust streaming off the nucleus.

Fast-forward to 2004. On 2 March, ESA launched its most daring and audacious mission ever – to send a spacecraft to rendezvous with a comet, go into orbit about it, and land a probe on its surface.

The *Rosetta* craft is named after the fabled Rosetta Stone, whose decoded hieroglyphs enabled archaeologists to understand the origins and development of the Egyptian civilisation. Scientists hope that the cosmic *Rosetta* will pull off a similar feat: by studying an ancient comet, we will learn of our Solar System's origins, and even about the development of life on Earth.

Its target? The wonderfully named Comet Churyumov-Gerasimenko. The cosmic interloper was discovered by Klim Ivanovych Churyumov in September 1969. At the observatory of the University of Kiev, he was studying a photographic plate from a Russian colleague – Svetlana Ivanova Gerasimenko, at the Alma-Ata Astrophysical Institute.

He rapidly realised that a fuzzy blob on Gerasimenko's plate was a new comet – and the pair now take the credit for discovering it. It swings around the Sun every 6.44 years, which means that it qualifies as a 'periodic comet': one that appears regularly. Its other name is Comet 67P: the 67th comet known to make a predictable assignment with the Sun.

This comet was not the first choice of ESA's space scientists. As flamboyant Project Scientist Matt Taylor explains: '67P was a back-up: originally we were targeted to go to 46P Wirtanen, but a launcher issue meant that we had a delay – and, very quickly, an alternative target was selected. It had to be periodic, and in a good place for us to get to. We had the spacecraft already built, so we couldn't modify it too much. In the end, we went for 67P; which, in hindsight, was an excellent choice!'

Taylor trained as a plasma physicist: 'I looked at things you can't see, like the solar wind and its interaction with the Earth's magnetic field – so the science bit behind the wonderful aurora.' He recalls: 'I was given an opportunity to work on *Rosetta* in 2013. It was a daunting one, but obviously a career highlight. I jumped at the chance, and have been waxing lyrical about the attractions of cometary bodies and *Rosetta* ever since.'

Rosetta's journey to its target comet was highly convoluted. In order to get up-close and personal to a comet – and gently fly alongside it – the probe had to undertake a complex series of interplanetary manoeuvres. In its ten-year mission to reach the comet, *Rosetta* had to swing past the Earth – three times – and Mars in order to match its speed to that of its celestial quarry.

Matt Taylor is astonished by this remarkable celestial achievement, which was all down to ESA's Space Dynamics Team.

'It's incredible to think that this ten-year-plus journey was choreographed from the get-go. We needed the planets to get us out to the comet's orbit and do the rendezvous. That gave us extra science, too. It's important not to forget that we also targeted two asteroids on the way. Mental!'

As *Rosetta* homed in on Comet 67P, its cameras were sending home amazing images of the comet, which looked like a black rubber duck, coated in dark ice-floes. Slipping into orbit around the comet, *Rosetta*'s instruments revealed that it smelled of rotten eggs and horse manure, coupled with the acid-sharp tang of sulphur dioxide. The putrid stench was only alleviated by the tang of bitter almonds – a telltale sign of deadly cyanide, mixed with scentless but poisonous carbon monoxide.

Disappointed? Want to go home? Not so: Taylor and his fellow scientists were elated. For this pungent mix of chemicals – sniffed out for the first time at Comet 67P – is the very elixir of life. It's probably the original material from which all living things have been made.

Then came the astonishing highlight of *Rosetta*'s mission: to soft-land a probe on the comet's dark, smelly surface.

During a nerve-wracking few hours on 12 November 2014, *Philae* – *Rosetta*'s tiny probe, no bigger than a dishwasher – began its epic descent to the surface of 67P. The Flight Dynamics Team had calculated its trajectory and velocity with impeccable precision. It was travelling at one metre per second – walking pace – so that it would have a gentle landing.

In the event, things didn't turn out as planned. *Philae* landed – bounced – rose a kilometre – bounced again – and finally settled down.

'We knew that landing would be tough,' reflects Matt Taylor. 'The gravity is 100,000th of that on Earth. We had implemented measures to stop the bounce – a thruster on the top of the lander to push it down, plus a couple of harpoons – but these didn't work.'

How did Taylor and his team feel about *Philae*'s landing, shortly after the event itself?

'It was a mix of relief and excitement. To be honest, I don't think it has sunk in. I spent all week doing interviews, and now I'm back in the office getting on with the complexities of the mission, so I haven't had time to think about it. I think that's for Christmas, over beers.'

Philae's final landing place was in a crater, whose mighty cliffs blocked out almost all the sunlight which would charge its solar panels. The probe had to rely on battery power – and that ran out in only two days.

Philae went into sleep mode. Before she entered her slumber, she tweeted: 'I'm feeling a bit tired. Did you get all my data? I might take a nap …'

As of now (February 2015), comet scientists don't know the exact location of *Philae* on 67P's surface. Matt Taylor observes: 'Once we get an accurate fix, we'll be able to determine what possible illumination conditions we will have. We have to wait and see. Even if we don't get *Philae* back, we have over a year to spend at this lovely comet with *Rosetta*. It's going to be nuts.'

So – with many months to come orbiting a comet, what do Matt Taylor and his team at ESA hope to learn?

'I think it's back to understanding how a comet works,' Taylor muses. 'This then allows us to tap into the big picture of comets being a leftover from the early Solar System – a building block, giving us an insight into where we came from and how the planets evolved.

'I feel it has also been something of a mission. It's engaging, and people are interested. And it's amazing to see kids so interested in a geek subject! Working together internationally for a greater good and the pursuit of knowledge is engaging everyone. We're all along for the ride.'

Whatever *Philae*'s fate, *Rosetta* is still alive and well – and in orbit about Comet 67P. We ask Matt about a possibility of landing the space probe on the comet. So – while the comet is still active, and the space probe still functioning – is this a possibility?

He replies: 'This, for me, is a strong possibility, and I feel a wonderful and spectacular conclusion to this mission. But there is plenty of science to do yet!!'

Chapter 15

IMPACT!

It's 9.20 a.m., on the morning after Valentine's Day, 2013. And if the Earth hasn't moved for any young couples in the central Russian city of Chelyabinsk the previous night, it's about to do so right now ...

'My heart is still beating 200 heartbeats a minute!' one resident posts on a local web forum. 'Oh my God, I thought the war had begun.'

A double whammy has hit the city: 'I saw a huge bright fire light up the skies, and then came a loud explosion that not only shattered the windows, but blew out the window frames.'

'Furniture was jumping,' says another eye-witness. 'What do I do? I grabbed my cat and my passport, and ran outside.'

Chelyabinsk indeed looks as if it has been in the wars. Around 100,000 window panes are shattered, sending a deadly hail of glass shards onto the pavements. Factory walls and roofs have collapsed. More than a thousand people have been injured; miraculously, no one is killed.

Social media instantly flash the news around the world. And, thanks to the chaotic state of Russia's traffic system, the nature of the explosion is instantly revealed. Russian drivers commonly install a dash-cam that films through the windscreen, to prove their innocence in cases of accident scams.

The dash-cams show a brilliant light in the sky. Speeding from the direction of sunrise, it grows ever more brilliant until it flares up, far outshining the Sun itself. The fireball disappears over the horizon, leaving a long smoky trail slashed across the sky. Two minutes later, the city is smashed by a wall of sound from the exploding fireball.

An asteroid has hit the Earth.

Once it's clear they haven't witnessed the outbreak of World War III, the citizens of Chelyabinsk start to relax. One local wit posts on Twitter: 'Chelyabinsk guys are so tough that – when they make a Valentine Day's promise to bring down the stars from heaven for their girlfriends – the stars actually drop!'

What dropped that morning was actually a chunk of rock the size of a six-storey office block, weighing over 10,000 tonnes. The asteroid sped in from space twenty times faster than a rifle bullet. Reaching incandescence as it crashed into the Earth's atmosphere, the interplanetary projectile exploded in a blast thirty times more powerful than the atom bomb that destroyed Hiroshima in August 1945.

While astronomers have known about stars and planets for millennia, the idea of *rocks* in space is quite new: as recently as 1800, many serious scientists wouldn't countenance such a daft idea. Yes, 'uneducated' country people had reported stones falling from the sky; but scientists contended that they'd either been ejected by volcanoes, or had coalesced from the atmosphere, like hailstones.

'Everything happens in the air, within that immense laboratory where lightning, hail and storms gather,' wrote a contemporary French journalist. Decrying the extraterrestrial theory, he continued: 'One should not worry; we are not at war with the Moon, and it is our planet only that performs hostilities on Earth.'

But the universe was about to 'perform hostilities' on our planet. On the clear afternoon of 26 April 1803, a massive shower of stones fell near the town of L'Aigle, in Normandy. A young French scientist was sent to investigate.

Jean-Baptiste Biot was amazed by what he found. There were plenty of reliable witnesses. One stone fell on a pavement in a churchyard, bounced in the air, and landed at the feet of a highly surprised chaplain.

Biot had no doubt the stones fell from space, but he set about bolstering his case. He checked there weren't any volcanoes nearby; and he brought some of the local stone back to Paris, to prove the strange rocks weren't just caused by lightning hitting the ground.

When he read his paper to the Institut de France, Biot's evidence changed the world's thinking overnight. Just a month after its publication, a senior professor summed up the impact of Biot's report: 'Few facts are more established in physics than the fall of meteoritic stones. And within a few months, we moved from doubt to certainty.'

The message spread rapidly throughout the world, largely thanks to Biot's passion to communicate his science. Unlike many scientists of the time (and even some today!), Biot wasn't going to hide his results within the covers of an academic journal. He believed science writing should be on a par with serious literature. In Biot's own words: 'Without sciences, the most literate nation would become weak and soon enslaved; without letters the most knowledgeable nation would fall back into barbarity.'

The internet – rather than fine literature – promoted the Chelyabinsk meteorite to the world's headlines in 2013. As well as the eye-witness reports and dash-cam images of the fireball passing overhead, a security camera showed a plume of ice and snow erupting as the major surviving fragment of the meteorite plummeted into a nearby frozen lake. And, after the ice melted, divers recovered a half-tonne chunk of the cosmic projectile.

A meteorite fall as spectacular as Chelyabinsk is a once-in-a-century event. But every year there are reports of smaller meteorites impacting the Earth. Fortunately, falling rocks from space are not as dangerous as they may seem: humans are fairly thinly spread across the globe, and often we are indoors.

Even so, a few people have been hit by meteorites. The best-recorded case occurred in 1954. A woman was snoozing on her sofa in Sylacauga, Alabama, when a meteorite the size of a grapefruit crashed through her roof. It destroyed her radio, then bounced off and hit her on the hip. Photographs taken in hospital show serious bruises down her left side.

In 1992, a student in New York had a more fortunate encounter. Unknown to Michelle Knapp, a meteorite had been filmed flashing across the skies of Pennsylvania and New York. The first inkling she had was a loud crash in her driveway. She suspected thieves or vandals; but she found the back of her old Chevrolet Malibu had been crushed by a space-rock.

Her insurance company refused to pay up, saying it had been 'an act of God'. Knapp, however, had the last laugh, by selling both the meteorite and her uniquely reshaped car for a fortune!

But the bulk of all meteorites garnered by scientists from across the world haven't been seen to fall. They've been discovered on the ground – hundreds or even thousands of years after they made their fiery descent.

The Inuit inhabitants of northern Greenland venerated a massive chunk of dark iron which they called 'The Tent' – a meteorite that fell to Earth around 10,000 years ago. In the late nineteenth century, the American arctic explorer Robert Peary tracked down this fabled piece of extraterrestrial real estate. To move the 31-tonne ingot, he had to construct a special railway through the snow, then ship it to New York – where you can still see it in the American Museum of Natural History, labelled as the Cape York meteorite. What science gained the Inuit lost – a valuable source of iron for their harpoons and knives.

The world's most venerated meteorite is worshipped every year by millions of pilgrims, travelling to Mecca. In the wall of the central sacred building, the Kaaba, is a black stone set in a silver frame. According to tradition, God sent this holy rock from the sky to show Adam and Eve where to erect their first altar.

The story clearly speaks of a meteorite; but scientists aren't allowed to take any fragments of the Black Stone for analysis. There are plenty of ideas: an agate, a piece of basalt or even black pumice. But the recent discovery of debris from a meteorite impact in the region suggests that the original idea of a divine messenger may be closest to the truth.

The deserts of the Middle East and North Africa are certainly a fertile hunting ground for meteorites. In this arid climate, even a fairly fragile stone from space won't be eroded away too quickly. And meteorites, blackened by their blazing path through the atmosphere, stand out from the red desert background.

Scientific expeditions to search for extraterrestrial stones here were spearheaded by British scientist Colin Pillinger in the 1980s. Pillinger was later to become famous as the 'father' of the *Beagle 2* probe that landed on Mars in 2003 but failed when its solar panels didn't open properly – as revealed by images from an orbiting NASA probe early in 2015.

But the scientific hunt turned into the 'Sahara Gold Rush' in the 1990s, when local people realised how much money they could make by selling meteorites to private collectors. Local nomads learned how to recognise meteorites, and sold them on to dealers, who traded the stones on eBay.

Thousands of meteorites emerged from the deserts – such a glut that the smaller stones weren't sold individually, but 'by the pound' – like potatoes. The buyers hoped that somewhere in this jumble they'd find a truly rare specimen: the rarest meteorites are literally worth their weight in gold.

Away from these commercial pressures, scientists have been turning to the pristine ice sheets of Antarctica. Like a conveyor belt, the moving ice sweeps any fallen meteorites along with the flow. If its motion is stopped by a mountain range, the top layers of ice gradually evaporate, to reveal the meteorites that were embedded within the ice sheet. They are easy to pick out, as dark lumps silhouetted against the dazzling frozen surface.

The most unusual of the Antarctic meteorites are stones that have travelled to Earth from the Moon, or from Mars. A Martian meteorite, found in 1984, achieved fame – or notoriety – twelve years later when

NASA scientists claimed that it contained fossilised microbes from the Red Planet. 'Life on Mars – Official', screamed the newspaper headlines.

President Clinton gushed: 'Rock ALH84001 speaks to us across all those billions of years and millions of miles ... If this discovery is confirmed, it will surely be one of the most stunning insights into our universe that science has ever uncovered.'

Today, most scientists have discounted this evidence, doubting that the 'wormlike' structures in the rock are really fossils. But even the most mundane meteorites provide us with stunning insights about our origins. Ironically, it's not the rare meteorites from Mars that scientists study for this deep information. The message-in-a-bottle is carried by the most common meteorites with their homeland beyond the orbit of Mars: the Asteroid Belt.

By an amazing cosmic coincidence, the first asteroids were discovered at the same time as Jean-Baptiste Biot proved that stones fall from space – though it was a long time before astronomers made the connection.

In January 1801, the Italian astronomer Giuseppe Piazzi was combing the skies from his observatory in Sicily, when he stumbled across an unexpected 'star' in the constellation Taurus.

'Cataloguing star positions was then the routine of the vast majority of astronomers,' explains David Hughes, a British expert on the small bodies of our Solar System. 'They weren't astrophysicists; they weren't interested in the composition of the star, they were just interested in where they were.'

Intrigued, Piazzi checked out the 'star' again the following night. To his surprise, it had moved ... For astronomers at that time, celestial wanderers were almost certain to be comets. Twenty years earlier, William Herschel thought he'd stumbled over a comet when he'd actually found the planet Uranus.

In public, Piazzi was suitably cautious – but he was brimming over with excitement. He wrote to a friend: 'I have announced this star as a comet, but since it is not accompanied by any nebulosity ... it has occurred to me several times that it might be something better than a comet.'

As astronomers followed the new object, and scrutinised it under high magnification, they realised that Piazzi had indeed stumbled on 'something better than a comet'. It was a totally new kind of astronomical body – a lump of rock less than a thousand kilometres across, circling the Sun between the orbits of Mars and Jupiter. Piazzi named it Ceres, after the patron goddess of Sicily.

And Ceres was not alone. Over the next few years, a consortium of planet-hunters – the self-styled Celestial Police – tracked down and apprehended three more of these worlds. But what to call them? As they are far smaller than the planets of the Solar System, some astronomers referred to them as 'minor planets'. Others homed in on their appearance as seen through even the most powerful telescope: the term 'asteroid' (star-like) is now most commonly used.

'So, within the course of one long generation,' NASA asteroid expert Steve Ostro explained to us before his sad death in 2008, 'we went from thinking of the Solar System as the Sun and six planets with assorted moons, to a Solar System containing comets and asteroids and also some little objects that fell to the Earth.

'As time has progressed,' he continued, 'space seemingly has got fuller and fuller, and we realise that all these minor bodies are moving around the whole Solar System like a swarm of bees.'

With the latest electronically equipped telescopes, astronomers have discovered over half a million asteroids. And that's a huge family to devise names for. Every asteroid has a number, for a start, linked to a vast computer database with all the information that's known about its orbit, its brightness and its nature.

But astronomers have carried on Piazzi's tradition of giving these little worlds real names. After Ceres, the next three – Pallas, Juno and Vesta – also commemorated goddesses. When the classical pantheon ran dry, the asteroid zoo began to include countries, cities and famous people from the past – sometimes with an extra '-a' on the end to make them appear feminine.

Don Yeomans, who studies asteroids and comets on behalf of NASA, points out that astronomers can use asteroid numbers and their linked names to invent 'coded' messages. For instance, if you replace asteroid numbers 9007, 673, 449, 848 and 1136 with their names, you find 'James Bond Edda Hamburga Inna Mercedes'!

Originally, anyone who discovered an asteroid had the right to name it. But astronomers became wary when they found minor planets named after a cat, three dogs – and the discoverer's mistress. Now the names are vetted by the International Astronomical Union, and the authors of this book are highly honoured to have their own celestial real estate, in the form of 3795 Nigel and 3922 Heather.

We've no idea what our celestial counterparts look like, but spacecraft have flown past a dozen asteroids, and sent back intimate shots. A NASA mission has landed on asteroid Eros; while a Japanese space probe extracted some dust from Itokawa and returned it to Earth for a proper lab examination.

It may seem an inordinate amount of attention to be directing to the puniest objects in the Solar System; but there's good reason for this intense scrutiny. Asteroids are builder's rubble left over from the birth of the Solar System: they provide the key to understanding the origin of our planet Earth.

'Four-and-a-half billion years ago,' David Hughes elucidates, 'the Sun was surrounded by a flat disc, containing dust and water vapour. The dust condensed to form rocky lumps, and out of this you made Mercury, Venus, Earth, Mars. You're also trying to form another planet where the Asteroid Belt is.'

But mighty Jupiter was growing in the background. 'Instead of this material coming together to form an individual planet,' Hughes continues, 'Jupiter's gravity perturbed them so they started smashing each other up. That's why – between Mars and Jupiter – we have literally a collection of bits.'

By investigating the bigger asteroids, we are seeing the building blocks of the planets – known in the trade as planetesimals. And the smaller asteroids – the broken-up remains of these small worlds – are providing an inside view of planetesimals.

The most ambitious mission to the asteroids is the twin-target *Dawn* orbiter; before moving on to Ceres in 2015, *Dawn* spent a year circling Vesta. The spacecraft found that this asteroid is far more than a simple rock. It's big enough that its centre melted, to form an iron core – like the Earth's central regions. Its surface has a complex history, too: Vesta is largely covered with old lava flows, showing that this small world was once a seething hell of volcanic activity.

Now, Vesta's face is quiescent – but it's scarred by a vast crater, almost as wide as the asteroid itself, with a central peak that's the highest mountain in the Solar System. The crater is named after a legendary Vestal Virgin, Rhea Silvia, who – despite her occupation – gave birth to the founders of Rome, Romulus and Remus.

The cosmic impact that created the Rheasilvia crater blasted trillions of tonnes of rock out into space. And some of these fragments have travelled across interplanetary space – as the astonished inhabitants of Johnstown, Colorado, discovered in 1924.

A series of explosions, like machine-gun fire, erupted over a baseball game. According to a local newspaper report, the spectators looked up to see 'the visitor whirling thru space, apparently headed for the center of the diamond. The game broke up in a stampede.'

The meteorite missed the baseball ground, and crashed to Earth outside the local church, where a funeral was being held. After the service, the undertaker disinterred a piece of blackened rock that was – according to the contemporary report – 'the size and shape of a man's head'.

Geologists have discovered that the Johnstown meteorite is – very unusually – made of solidified lava. It's an exact match to the volcanic rocks that the *Dawn* mission found covering the surface of Vesta.

The smaller asteroids were never as volcanic as Vesta, and their dark surfaces are coated with rocks that have hardly changed since the birth of the Solar

System. When one of these fragments falls to Earth, the scientists really jump!

In January 2000, the inhabitants of a remote region of the Yukon, in western Canada, had a rude awakening when a fragment from the almost-black asteroid Irmintraud smashed into the Earth's atmosphere. According to local guide Jim Brook, the morning sky was lit up by a brilliant fireball that 'took several seconds to move right across the sky. Then there was this massive explosion: I could feel myself shake with it.'

A week later, as he drove home across frozen Lake Tagish, Brook 'came round the corner and found these lumps of rock – smelling of sulphur'. Suspecting that the walnut-sized specimens could be precious, Brook picked them up with plastic covering his fingers – to prevent contamination – and stored them in his fridge until he could hand them over to scientists.

And the Tagish Lake meteorite was a cosmic bonanza. The crumbly black rock was rich with compounds containing carbon – the raw material of life. Scientists were fortunate that Brook had kept the samples so pure; and they were lucky that any solid pieces of this fragile rock had reached the ground at all.

'The Tagish Lake meteorite was a 150-tonne object,' comments Don Brownlee, an American astronomer who studied the rocks that Brook had picked up. 'As it punched its way through the atmosphere, the whole thing just blew up. It wasn't strong enough to survive as big pieces, and the original meteorite was mostly busted up into dust.'

The fragile Tagish Lake rocks were never heated or compressed. They must have resided on the surface of asteroid Irmintraud, unaltered since the birth of the

Solar System. By checking out the different radioactive atoms in the rock, scientists could measure its age – just as archaeologists have used radio-carbon to discover when Stonehenge was built.

The result comes in at 4.6 billion years. That's how long ago the Sun and the planets were born.

At the opposite extreme to the fragile Tagish Lake meteorite are solid lumps of metal that can crash through the atmosphere without breaking up, like the massive Cape York meteorite that Robert Peary shipped out of Greenland. These 'iron' meteorites are actually a hard alloy of iron and nickel that have been smashed out of an asteroid's core. And, when it lands, a big iron meteorite can pack a mighty punch.

Road signs along the old Route 66 through northern Arizona used to make this point – perhaps not too subtly – by directing visitors to 'The planet's most penetrating Natural Attraction'.

Meteor Crater is a huge hole in the ground, more than a kilometre wide and almost 200 metres deep. It's big enough to swallow up fifty New York city blocks, and the crater is so deep at its centre that only the tallest skyscrapers would poke up above ground level. It was blasted out by a small iron asteroid that smashed into the Arizona desert around 50,000 years ago. The meteorite penetrated deep underground; then its enormous energy of motion converted to explosive power. It erupted with the power of more than a thousand Hiroshima bombs.

But even Meteor Crater is puny on the cosmic scale. Look at the Man in the Moon through binoculars – or a small telescope – and you'll see his face is pockmarked with craters that are far larger. It resembles a cosmic battlefield – because that's what it is. Countless projectiles from the asteroid belt have exploded on its surface.

Throughout Earth's history, our planet must have suffered similar bombardment. But we're not surrounded by craters because erosion has worn them away. Rivers and waves constantly sandpaper our planet smooth, while the mighty forces of shifting continents are pushing up new mountain ranges – not to mention volcanoes resurfacing the Earth. Only with the greatest diligence can geologists trace the remains of old craters lying below the Earth's current surface.

The record of cosmic violence can be read, however, in the fossils of past life on planet Earth. Animals and plants are sensitive souls, and any dramatic shock to their environment will lay them low – as the dinosaurs found, 66 million years ago …

These great animals had lorded it over the Earth for more than 100 million years. And then they suddenly disappeared. And it wasn't just the dinosaurs.

'Over 75 per cent of the species of life – plants and animals; on sea, on land – went extinct,' says American geologist David Kring. Only a few survivors of each species would be sufficient to recolonise the world, so the actual body count must have been far higher. Kring's best estimate is that 'probably on the order of 99.9999 per cent of the individual organisms that lived at that time were killed.'

Kring and his colleagues have proved that this global carnage was the result of a massive hammer blow from space, on a scale that makes the Meteor Crater impact look like a damp squib. Unlike Meteor Crater, this giant scar wasn't something that scientists were just going to stumble over. It's buried under a thousand metres of more recent rocks, on the edge of Mexico's Yucatan Peninsula.

Bizarrely, it was a team of oil prospectors that unwittingly located the smoking gun for the death of the dinosaurs. Around 1980, they discovered tiny disturbances in the Earth's magnetic field and gravity near the Mexican coastal town of Chicxulub, which suggested a vast ring-shaped structure half underground and half under the sea. They drilled deeply to check the geology; found no evidence for oil-containing rocks; and filed the reports safely away from the eyes of their commercial rivals.

Kring, meanwhile, was intrigued by the then-new theory that the dinosaurs had been killed by an asteroid hitting the Earth. Such an impact must have blasted out a vast crater, and Kring was determined to find it – by looking for the broken rocks flung out for vast distances. 'When impact ejecta is deposited, it is thicker towards the crater, and thinner away from the crater,' he explains. 'So you can go out into the world, and – like breadcrumbs – follow them back to the crater as you find thicker and thicker deposits.'

In the United States, Kring found a layer of broken rock laid down 66 million years ago. It became thicker as he followed it further south, into the Caribbean. His sights became focused on Mexico. One of his team heard of the oil prospecting there; and got hold of the rock samples drilled from deep underground. These cores provided the clinching evidence: they contained minerals that are created only in the fury of an asteroid strike. At exactly the time that the dinosaurs died, a giant space rock had hit Yucatan and blasted out the world's biggest impact crater.

'We're talking about the impact of an asteroid maybe ten to twenty kilometres across,' says Kring. If the dinosaurs had been looking upwards, 'it would have been visible three, four – maybe five – days before it actually hit the Earth'.

This was a space rock the size of Mount Everest, hurtling towards us at 100,000 kilometres per hour. It ploughed through the Earth's atmosphere in a matter of seconds. 'At that point, things began to happen very, very quickly,' Kring continues. 'We would have had an extraordinary sky-splitting, blinding type of fireball.'

Hitting the Earth's surface, it ploughed downwards – and then exploded. 'We're talking about an event,' says Kring, 'that had the explosive energy of one hundred trillion tonnes of TNT.'

The asteroid vaporised, and an intense shock wave raced outwards from the explosion through the Earth's crust, shattering and melting the rocks. 'It also depressed the surface of the Earth,' Kring explains, 'and then it began to rebound, with a tremendous amount of uplift. Rocks from the middle of the Earth's crust came to the surface; and were then raised into a huge mountain rising into the sky.'

But this majestic mountain was only temporary. 'The weight of this rising pillar of rock was actually too great to be supported, and then it came crashing down, excavating material in a crater.'

The impact raised giant tsunamis that overwhelmed the land around the Caribbean. It unleashed an airblast – a supersonic wind that destroyed all living things nearby. But these local effects were only the beginning …

'It's the global effects that were truly important to the alteration of biological evolution,' Kring emphasises.

Debris from the explosion was thrown far into space, travelling halfway towards the Moon before Earth's gravity started to pull it back down. The whole sky was full of rocks burning up as meteors: 'you can think of it as fiery rain,' says Kring, 'which enveloped the entire globe.'

The burning sky set fire to forests around the world. The dinosaurs – and other land-dwelling animals – were literally roasted to death.

The fiery furnace brewed up poisonous chemicals in the atmosphere, along with a layer of smog that threw the Earth into the freezer. As the sky cleared, carbon dioxide from the impact created an intense Greenhouse Effect that heated our planet for thousands of years afterwards.

The triple whammy of grilling, freezing and baking was enough to exterminate almost all life on our planet.

The age of the giant dinosaurs had passed. Only the smallest of the family survived, to become the birds of today. Some little furry mammals lived on, too, protected in burrows while the fiery rain fell. Their diet helped their survival: mammals are not fussy eaters, and could munch on whatever else had pulled through the firestorm, such as insects and aquatic plants – not to mention tasty titbits of toasted Tyrannosaurus …

With little competition on land, mammals flourished – and evolved into the huge diversity of wildlife we see today, from kangaroos to koalas, from elephants to elands, from hyraxes to humans.

And now it's we humans who are in the firing line. The fireball that erupted over Chelyabinsk early in 2013 was a wake-up call that there's still plenty of ammunition in the cosmic shooting gallery.

'These "near-Earth" asteroids cross the orbit of our planet, and they all have the potential to collide with the Earth one day,' said Steve Ostro. 'What we need to do is discover as many of these objects as we can – certainly all the really large objects, the kilometre and larger size ones.'

A handful of observatories around the world are devoting their time, not to distant stars and galaxies, but to these potential killers on our doorstep. Some are run by devoted amateurs working in their spare time. A professional surgeon based in Spain discovered an asteroid that whizzed closely past the Earth – by an amazing cosmic coincidence – on the very same day that the Chelyabinsk meteor struck.

'Modern amateur-scale equipment rivals the best professional equipment of only fifteen to twenty years ago,' applauds Steve Larson, the head of the Catalina Sky Survey, the world's largest search campaign. Based in Arizona, Larson's team found the first asteroid predicted to hit the Earth. In 2008, their telescope picked up a space rock the size of a lorry, just beyond the Moon's orbit. Within twenty-four hours, it would wallop our planet – in northern Sudan.

They alerted the Pentagon – and even the White House – who in turn contacted the Sudanese government. Their calculations showed it would impact in the uninhabited desert. Right on time, the pilot of a KLM airliner, flying over Sudan, witnessed a brilliant fireball.

The Catalina telescopes will soon be overtaken by a vast instrument in Chile that will sweep the sky for much fainter objects. The prospect is exciting NASA expert Don Yeomans, who has explained his mission in a book called *Near-Earth Objects: Finding Them Before They Find Us*: 'Once the Large Synoptic Survey Telescope comes on line, they will be the big dog and dominate the ground-based surveys.'

But telescopes on Earth can't find all the potential threats from space: they can only discover menacing objects that appear in the dark night sky – coming from beyond the Earth's orbit. They all missed the Chelyabinsk asteroid because – in the language of fighter pilots – it came 'out of the Sun'.

Zooming into our atmosphere from within the Earth's orbit, the Chelyabinsk asteroid was never visible in a dark sky. To pin down wayward marauders like this, we need a telescope that's in space, and positioned closer to the Sun, so it can look outwards to the Earth and any dangerous interlopers heading our way.

Enter Ed Lu. He knows a lot about space, having been an astronaut on two US Space Shuttle missions and an inhabitant of the International Space Station for six months. Lu is planning a space telescope, based near Venus, that will hunt out dangerous asteroids. And he's not worried about government dithering, and the capricious winds of politics, as the Sentinel telescope will be funded by members of the public. 'Private organisations can now carry out awe-inspiring and audacious missions that previously only governments could accomplish,' he says. 'We are conducting Sentinel in the private sector because we can!'

Yeomans is admiring: 'If funding can be secured, Sentinel will be the major system for discovering near-Earth objects. And – working at infrared wavelengths – they will be able to provide much better constraints on the sizes of the objects they find.'

Size is everything when it comes to impacts; and how to cope. If astronomers know that a small rock is on its way – say the size of the asteroid that fell near Chelyabinsk or in Sudan in 2008 – it would be best to treat the threat in the same way as a predicted hurricane or volcanic eruption. 'Policy makers would have to decide,' says Yeomans, 'whether to take the hit (it would more than likely impact harmlessly in an ocean or unpopulated area); or whether to evacuate the threatened area.'

Coming to asteroids around one kilometre across, Yeomans reassures us: 'We've discovered about 94 per cent of the largest near-Earth asteroids, and none of these represent a credible threat for the next hundred years.'

Everest-sized asteroids are fewer and far between. And a monster space rock doesn't have to be a planet-wide killer – provided we have enough advance warning that it's on its way. If we have a century's notice, we can dodge the bullet – not by moving the Earth, but by deflecting the asteroid.

'With our present state of space development,' maintains David Hughes, 'we don't just have to sit back and take it.' There's no need to send Bruce Willis, as in the 1998 Hollywood blockbuster *Armageddon*. Robotic spacecraft are every bit as capable.

First we need to know the asteroid's flight path with precision. Steve Ostro regularly used giant radio telescopes like an airport radar system to track the flying rocks. 'The most magical experience is getting the first echo,' he enthused, 'then using the full power of the radar astronomy technique to see images – it's almost like the *Star Trek* experience.

'It's a momentous experience emotionally,' he continued, 'because we know that – throughout the rest of the history of humanity – these objects are going to play a big role.' Not only could Ostro plot the path of the asteroid, he could also work out what it's made of.

And that's important for our cosmic protection. Suppose we try to deflect an asteroid by hitting it with a high-speed spacecraft or a nuclear blast, and it's a fragile rock that simply shatters and carries on its original course. The Earth will still be hit – not by a cosmic rifle bullet, but by an equally powerful shotgun blast.

Astronaut Ed Lu recommends a 'gravity tractor'. Instead of touching the asteroid at all, we would put a spacecraft nearby, so the two are tethered by bonds of gravity. We gently fire rockets on the spacecraft; and the gravity pulls the asteroid along as well, like a dog on an invisible leash.

'We can do something about the asteroid threat nowadays,' says David Hughes, 'and I think we are therefore duty bound to do something about it.'

Our prowess in space means that – for the first time in the planet's history – the inhabitants of planet Earth can protect terrestrial life from wholesale extinction. And it will alter the course of evolution on our planet.

The death of the dinosaurs wasn't the only large-scale massacre of life on our planet – just the latest. During the past half-billion years there've been five major mass extinctions – though it's hard to study the earlier disasters because the rocks and fossils have been altered over vast reaches of time.

Kring is convinced that asteroid impacts were responsible. 'From the statistics of impact craters, we know there should have been five to six events the size of Chicxulub during that time,' he reasons. 'So if

Chicxulub is a measure of what could happen to the environment, it seems likely there should be similar points in time where mass extinctions occurred.'

'These impacts have devastated various geographical areas, and in the largest energy extreme have eradicated most of life on the planet,' concurred Steve Ostro. 'They effectively reset the evolutionary clock, irrevocably altering the direction that evolution was taking.'

And that wasn't necessarily a bad thing. When the climate on Earth was stable, life just carried on the status quo. That's why the dinosaurs ruled unchallenged for so long.

'Asteroid impacts played an active role in changing the course of evolution,' said Ostro, 'opening up new ecological niches. We are the current end products.'

The mayhem of an asteroid impact gave evolution a huge push forward. Five times over, life on Earth has been kick-started to a new level: *Homo sapiens* and our civilisation are the latest benefactors. Without the cataclysms of asteroid attack, we would not exist. As Don Brownlee puts it, 'If the dinosaurs hadn't been wiped out by an impact, we wouldn't be standing here: it would still be the age of the reptiles.'

Chapter 16

IS THERE LIFE ON MARS?

'Ladies and gentlemen, I have a grave announcement to make.' With these dire words, a continuity announcer interrupted a CBS radio programme in October 1938. He continued: 'The strange object that fell at Grover's Mill, New Jersey, earlier this evening was not a meteorite. Incredible as it seems, it contained strange beings who are believed to be the vanguard of an army from the planet Mars.'

As it turned out, the 'continuity announcer' was the 23-year-old Orson Welles – a brilliant actor and director who had previously been a bullfighter and magician. He was dramatising H.G. Wells's 1898 novel *The War of the Worlds* as a live presentation for radio, relocated to America. Welles's awesome adaption convinced and terrified the American nation into believing that they were under threat from an alien invasion.

The broadcast was a stunt to improve audience ratings for CBS, which was losing listeners to a rival network. If Welles failed, his days in radio would be numbered. On that particular night, the gods were not just smiling on him – they were grinning deliriously. The other radio channel was featuring an unknown singer, and bored listeners started twiddling their dials ...

Welles certainly knew how to wind his listeners up. He had a 'reporter in the field' describe an alien emerging from its spacecraft. 'There, I can see the thing's body. It glistens like wet leather. But that face. It ... it's indescribable. The mouth is V-shaped with saliva dripping from its rimless lips that seem to quiver and pulsate ...'

A 'voice from Washington' revealed that the Martians were landing all over the United States. Apparently, thousands of people had already been killed in cold blood by the aliens' death-ray guns. The broadcast culminated with the Martians sweeping through Manhattan and invading the radio station itself. It ended with a chilling, high-pitched scream.

The acting was so convincing that many Americans fled their homes. They hit the streets, hid in cellars, prayed, loaded up their guns and even wrapped wet towels around their heads as protection from Martian poison gas. Some even claimed to have *seen* Martians.

Welles was blissfully unaware of what he had precipitated until he bought a newspaper the following morning. 'Radio listeners in panic,' screamed the headline. 'Many flee homes to escape gas raid from Mars.' People even brought lawsuits against CBS, but all were eventually withdrawn.

However, there's no such thing as bad publicity. Welles's career took off instantly, and CBS was delighted to have America's most notorious actor broadcasting on its airwaves. In an ironic twist, Welles was on radio just three years later on 7 December 1941 – the day of the Japanese attack on Pearl Harbor. When the (real) continuity announcer broke the news of the bombing, many of the American public believed that they had been hoodwinked again!

It was the latest episode in Mars's long association with war. Around 1000 BC, the Chaldeans, inhabitants of what is now Iraq, called the planet Nergal. He was a great hero: the king of conflicts, master of battles, and the champion of gods. The Greeks named the planet Ares, after a son of chief god Zeus. According to Homer's *Iliad*, Ares was disliked by many of the gods – and the Greeks themselves – because he was tempestuous and murderous. But to the Romans, who glorified conflict, Mars was their revered god of war: and it was they who gave the planet the name we use today.

For nearly 2,000 years, Mars remained a red dot in the sky, slowly moving against the background of stars. It wasn't until astronomers pointed the newly invented telescope to the heavens that they could see it was a real world. In 1659, the great Dutch astronomer Christiaan Huygens mapped a large dark spot on Mars, probably the region we now call Syrtis Major. By logging its rotation, Huygens found that the Red Planet has a 24-hour day.

Way in advance of his time, Huygens penned *Cosmothereos*, a 1698 book on how planets might sustain life. With the full title – in English – *Celestial Worlds Discover'd: or, Conjectures Concerning the Inhabitants, Plants and Productions of the Worlds in the Planets,* this volume was an international publishing sensation. And it was to be the beginning of something very, very big ...

In 1877, Mars and the Earth drew particularly close together in their paths around the Sun. In Milan, Giovanni Schiaparelli observed that the Red Planet was crisscrossed with a dense network of overlapping lines – which he called 'canali', or 'channels'.

But the word crossed the Atlantic, and it suffered in mistranslation. 'Channels' became canals: implying constructions of artificial origin. In particular, it reached the ears of Percival Lowell.

We've already introduced Lowell in Chapter 7, as the wealthy and learned Boston businessman who built the world's first high-altitude observatory in Flagstaff, Arizona, where he searched for a planet beyond Neptune. But his first obsession was with life on Mars.

Lowell was convinced that Schiaparelli's 'canali' were deliberately constructed feats of engineering built to transport water from the Martian poles to the desiccated deserts at its equator. Mars was a dying planet, he maintained. 'If the planet possesses inhabitants, there is but one course open to them to support life. Irrigation must be the all-engrossing Martian pursuit ... it must be the chief material concern of their lives,' he wrote.

At Flagstaff, Lowell sketched hundreds of straight canals. A committed pacifist, Lowell saw the global scheme as evidence that the Martians were united, and free from the scourge of war. Moreover, he believed that they were much more advanced than us: 'A mind of no mean order would seem to have presided over the system we see – a mind certainly of considerably more comprehensiveness than that which presides over the various departments of our own public works. Quite possibly, such Martian-folk are possessed of inventions of which we have not dreamed, and with them electrophones and kinetoscopes are things of a bygone past, preserved with veneration in museums as relics of the clumsy contrivances of the simple childhood of the race.'

Lowell was not one to keep his views to himself. As a result, what had been a demure astronomical debate about the nature of the canals turned into a raging public controversy. Even the severe economic depression of 1907 took its place on the back burner as compared to the Martians. That year, the *Wall Street Journal* asked its readers what had been the most extraordinary event of the past twelve months. The response was unequivocal: 'the proof afforded by astronomical observations … that conscious, intelligent human life exists upon the planet Mars'.

Martians were now well and truly in the public arena.

But were there really canals on Mars? With hindsight, we can see the fatal errors that were stacking up in the way that astronomers observed Mars. Small telescopes would show canals; large telescopes didn't: that was pointing to something odd going on. And now that space probes have surveyed the surface of Mars even more thoroughly than that of the Earth, we know that the canals don't exist. They were a by-product of astronomers straining their eyes through old-fashioned telescopes in days long before the electronic and digital revolution.

Neverthless: Mars is tantalisingly Earthlike. Although only just over half the size of our planet, it has polar caps, seasons, and boasts a day-length just a little longer than that of Earth. Its surface is dotted with dark markings, which – until as recently as the 1950s – were thought by some astronomers to be patches of vegetation.

But hopes for life on Mars were dashed in the summer of 1965 when NASA's *Mariner 4* probe flew past the planet.

It returned pictures of a desolate, cratered world much more like the Moon than the Earth. For a generation of scientists raised on the notion that Mars was similar to our own world, the images were a bitter blow.

Then, in 1971, along came *Mariner 9*. The first spacecraft to go into orbit around the Red Planet, *Mariner 9* was scheduled to make a one-year detailed photographic survey of Mars. But things didn't quite turn out that way. When the probe arrived, Mars was enveloped in a planet-wide dust storm, which had completely obliterated the surface.

Scientists waited impatiently for the pall of dust to clear. As it did, they began to pick out new features on the Red Planet: four dark spots, in particular. When the dust settled, the nature of the spots became clear. They are, in fact, huge volcanic cones whose tops have caved in – like the volcanoes on Hawaii. The largest – appropriately called Olympus Mons – towers twenty-nine kilometres above the surface. It is three times the height of Mount Everest, and wide enough to cover the whole of Spain.

Amazingly, the small planet was displaying volcanism on a scale much grander than the Earth. And *Mariner 9* went on to discover a vast canyon system – now called Valles Marineris in its honour – which was so large it would engulf the whole of the Alps. The orbiting space probe had turned around the reputation of the Red Planet from a dead world into one that had experienced an extraordinarily rich and diverse history.

Even as *Mariner 9* was circling Mars, plans were afoot for an even more ambitious mission to the Red Planet: the twin landers *Viking 1* and *Viking 2*. Their goal? To search for life.

In the summer of 1975, the *Viking* probes blasted off from Cape Canaveral. Each comprised an orbiter to circle Mars and survey its features, plus a lander to touch down gently on the surface.

Viking 1 was scheduled to land in the desert of Chryse Planitia on the auspicious date of 4 July 1976 – the US Bicentennial. But Mars refused to co-operate with the American Dream. The sharp eye of the new Orbiter showed that the landing site – selected from *Mariner 9* images – was anything but smooth, and much too dangerous to land on.

Mike Carr, head of the *Viking Orbiter* team, remembers vividly when the spacecraft's camera revealed the true extent of the site's roughness. 'It was really scary. We could see all this incredible detail. My God, I thought – we can't land there. There are so many obstacles in the field.'

The team started an intense search for a new landing site. 'I had this army of people doing terrain assessments, doing the geology,' recalls Carr. 'Every day, we'd get these new pictures coming in and we'd have a meeting to assess the new data, presided over in this big room by the Project Manager.'

'We'd put these wet prints up on the board and I had to interpret them – interpret the geology. I'd seen them only thirty seconds before, and I had to tell everyone what was there in the images. This went on for a few weeks, and we finally decided on the landing site we ultimately went to.'

In Mission Control, the science teams braced themselves for the first images from the *Viking 1* Lander – now safely on the surface 740 kilometres from the original site, and some 16 days late in touching down.

Would the technology work? 'It was tense,' recalls Carr. 'But then there was the excitement of seeing the first pictures. Lots of new information was coming in, and the excitement was very real.'

The Lander revealed a dusty-pink world with salmon-pink skies – a result of dust particles blown into the thin Martian atmosphere. But there was nothing rosy about the Red Planet. It was bitterly cold, and almost airless. Drifts of fine Martian soil stretched for miles, as powdery as snow in the Antarctic (where some of the Landers' tests had been carried out). Rocks and boulders of all shapes and sizes littered the scene. Many were rough and volcanic in appearance, with small holes where gas had once bubbled through – like pumice.

Viking 2 arrived at Mars a couple of months after its twin, and the Lander touched down at a gently sloping site called Utopia. Both Landers returned weekly weather reports, analyses of the Martian atmosphere, wind-speed readings, and thousands of pictures of the surface of Mars in all its moods. The *Viking 1* Lander lasted longest – a venerable six years – and after shutting down, it was renamed the 'Thomas A. Mutch Memorial Station', after a brilliant young *Viking* team-leader who was lost when climbing in the Himalayas.

Beyond the stunning *Viking* images, what captured the world's imagination were the experiments designed to detect life on Mars. The Landers each carried a miniature laboratory, the size of a wastepaper basket, to perform the life-detection experiments by remote control. Nothing as sophisticated had ever flown to another world.

Each *Viking* Lander scooped up samples of soil, and fed them into its laboratory. Here, three experiments tested for any activity in the soil, either chemical or – the Holy Grail! – biological.

Two of the experiments – one testing for plant-like life, and the other looking for gas given off by living cells – gave negative and inconclusive results, respectively. Any reactions that took place, agreed the researchers, were down to chemistry rather than biology.

But the third set of results had everyone sitting up, and taking a considerable amount of notice. They came from the 'Labelled Release' experiment. Its designer, Gil Levin, explains: 'The standard method of culturing micro-organisms is to put them in some kind of nutrient soup, and wait several days until they start multiplying and you can see them. My technique simply added radioisotopes to those nutrient compounds.'

If there were any bugs in the Martian soil, they would start burping out radioactive gas as soon as they began chewing through Levin's special nutrient. 'This would be detected much more quickly than waiting around for a visible gas bubble,' he continues. 'So the whole thing reduced about two days of waiting for evidence of life to about fifteen or thirty minutes.'

Mike Carr recalls the feeling in the Control Room when Levin's experiment yielded up copious quantities of radioactive gas. 'We initially thought, my God, my God – there may be life there. And then, it all kind of waned.'

The excitement about life in the Martian soil dwindled because NASA had a fourth experiment on each Lander. Ironically, it wasn't intended to look for life on Mars, but just to check out what the soil was made of. This instrument – the GCMS (if you'd like it in full, that's the gas chromatograph mass spectrometer) – didn't find any sign of carbon in the soil.

All life is made from carbon atoms. If there had been any micro-organisms in the Martian soil, then the GCMS should have sniffed out the carbon they're made of. Taking all the results together, NASA's official line on the Labelled Release experiment is that it didn't discover biology, but some kind of exotic chemistry – conveniently dismissing the fact that the GCMS was so insensitive that it would have missed more than 100 million microbes in a teaspoonful of soil.

Levin absolutely refused to take this assessment lying down – and we wanted to know why. We tracked him down to a light industrial estate just off the Beltway in Washington. At first he was – understandably – rather guarded. But when he realised that we were astrophysicists before going off to work in the media, he began to feel he could trust us.

Levin took some papers out of a filing cabinet, and placed a computer print-out on the table. 'Here's the data. As you can see, it continued to build at a strong rate until the second day, but continued even beyond that for the full eight days of the programme.'

To our eyes, the Mars data did not look like a chemical reaction. Chemical activity peaks rapidly and dies away. But the Martian curve builds steadily. Levin pointed to the next set of curves, from an identical test that's been done on Earth: 'Here are the controls that were run, so that we have signals from terrestrial soils with living micro-organisms.' It's certainly difficult to see any difference between the Martian results and the burps of radioactive gas from genuine living cells in his lab.

Levin had one last-ditch attempt to convince NASA that he was on to something. He decided to heat samples of Mars soil within a narrow range of temperatures –

covering the degree of heating that kills off bacteria on the Earth. 'First of all, we showed that 51 degrees definitely destroyed the signal. But secondly, we showed that 46 degrees didn't destroy it – it inhibited it by 30 per cent. And that's just the way that in the laboratory here we distinguish *E. coli* from the rest of the coliforms, because *E. coli* can survive beyond 37 degrees, while the others cannot.'

It was all to no avail. NASA had turned its collective back on Levin. 'It was political,' he acknowledges. 'They had to come down with a decision, and they hate to retract a decision. If you go to people from NASA and you say, what do you think of the Labelled Release experiment, they'll say: "Oh, that's garbage, you know. Levin keeps saying the same thing over and over again".'

But NASA has not entirely given up on Mars. Except that the agency refuses to acknowledge that it's searching for life – it is, in its own words, 'following the water'.

Water is the key to life. Our bodies are 60 per cent made of H_2O; without water, the chemical reactions that power life would grind to a halt. And in 2003, the agency launched two small rovers to the Red Planet, designed specifically to sniff out signs of past water on Mars. Both *Spirit* and *Opportunity* were meant to last for three months. *Spirit* lived for six years before getting stuck in soft soil, while *Opportunity* is still going strong. And – from Day One – both rovers confirmed that Mars is an active world.

Spirit landed inside Gusev Crater, where it spectacularly filmed dust devils skimming over the surface of the Red Planet. Of more scientific importance, its drilling tools revealed that the environment was the product of a volcanic explosion where magma and water had mixed together.

The rovers' principal investigator, Steve Squyres, enthusiastically describes his team's findings. He starts with *Spirit*'s adventures after driving off towards some neighbouring peaks: 'In the Columbia Hills, we discovered compelling evidence of an ancient Mars that was a hot, wet, violent place – with volcanic eruptions, hydrothermal activity, steam vents – nothing like Mars today. It had the key ingredients for life: liquid water and an energy source.'

Opportunity touched down on the opposite side of Mars – and instantly found signs of water flowing in the past. It detected 'blueberries': spherules of the greyish mineral haematite, which you may know from necklaces and bracelets that jewellers create back here on Earth. On Mars, the find was important because these concretions are created when waterborne minerals settle into sedimentary rock. And that was just the beginning of the rover's epic Mars Trek …

'When *Opportunity* got to the rim of Endeavour Crater,' says Squyres, 'we began a whole new mission. We found gypsum veins and a rich concentration of clay minerals. These tell us about water chemistry that was neutral, instead of acidic – more favourable for microbial life, if any ever began on Mars.'

Opportunity has now been joined by NASA's *Curiosity* Rover. The car-sized probe, which landed in 2012 in Gale Crater, is looking for any imprint that past life may have left on the Red Planet. And it's paving the way for a human expedition to Mars, which US President Barack Obama maintains could be possible within our lifetimes.

In a speech in April 2010, Obama predicted a manned conquest of Mars within a couple of decades: 'By the mid 2030s, I believe we can send humans to orbit Mars and return them safely to Earth. And a landing on Mars will follow. And I expect to be around to see it.'

A human mission to Mars will not be plain sailing. Crossing interplanetary space is not the same as the three-day hop that the Apollo astronauts made to the Moon. The mission will be bedevilled with a host of hazards. For a start, the travel time to Mars is a minimum of seven months – followed by eighteen months on the surface of the Red Planet before the orbits of the Earth and Mars get into register again. And then, there's the long journey home.

The dangers of the flight are almost incalculable. Humankind has never attempted anything as ambitious. Heading up the risk is radiation: in the unprotected seas of our storm-wracked Solar System, space weather will be a major hazard. The Sun, our gentle local star, can be a fierce beast – hurling out energetic charged particles when it goes through periodic violent phases during its eleven-year cycle of solar activity.

And more distant stars will have an effect on our future explorers. Exploding stars – supernovae – shoot deadly cosmic rays into space, and these heavy particles have lethal effects. Be prepared for some dead or severely injured astronauts. One NASA spokesman commented that 'we fully expect to get some wet noodles back'.

But the key problems will be psychological. Jack Stuster, senior scientist at Anacapa Science Incorporated in Santa Barbara, California, is looking into the implications of long-duration spaceflight. He's studied human performance right across the board

– from nuclear power plant technicians, through US Navy teams working under extreme conditions, to NASA astronauts.

Stuster likens a Martian mission to a nightmare holiday. 'Imagine travelling around England in a motor home with five other people for three years, and you can't go outside!'

There's not a lot of data on this particular kind of torment, so he also compares the flight to Mars with a long sea voyage. 'One of the effects of isolation and confinement,' he observes, 'is that trivial issues are exaggerated beyond all reasonable proportion. People blow up over some tiny conflict, and an hour or so later, they wonder what the hell went on.'

Despite all these caveats, nothing seems to have deterred the notion of mounting a human mission to Mars. The Americans, with their incredibly successful track record of exploring the Red Planet, are in the lead. But they're not the only ones. Russia, Europe and China have all proposed missions within NASA's timeframe.

And private space enterprise is muscling in – hoping to be ahead of their international competitors. The most publicised mission is Mars One, offering – literally – a one-way ticket to the Red Planet. It's run by a Dutch company seeking volunteers who are physically fit, psychologically sound, team players, and over the age of eighteen (there's no upper age limit!). And there's no coming back – because there isn't a spacecraft on Mars to return the crews to Earth.

The Mars One team is hoping to have earthlings living on the Red Planet as early as 2024. It sounds a tough call: the technology hasn't been tested, and the funding is not yet in place. But the basic concept is feasible with today's technology: sending rockets and rovers to Mars to build a habitat before the astronauts themselves arrive.

A permanent colony of humans would grow on the Red Planet, as groups of four crew members immigrate every two years (when Earth and Mars are at their closest). But the whole scheme does look a bit like a glorified cosmic caravan park. What might the putative settlers do? How will they explore the Red Planet? What will they do to while away their days in such an alien environment?

Yet this hasn't put adventurers off: more than 200,000 people applied to be astronauts, and they've been whittled down to 705 would-be Martians from 11 countries.

So: the dream is alive. However far-out the Mars One mission may seem, it's a sign that humanity's sights are firmly trained on the Red Planet. Even Jack Stuster, who knows more than anyone about the psychological and behavioural problems of sending a crew to Mars, would sign up for a visit to our neighbouring world. 'I'd go in a heartbeat – if I could take my family with me.'

To return to the big question: is there life on Mars? There will be; and in a shorter time than we expect. And it will be human life.

Chapter 17

NEW WORLDS FOR OLD

Singing in the shower is about the closest that most of us get to creativity during our morning ablutions; but the water droplets bouncing off American astronomer Geoff Marcy gave him an inspiration that changed not only his life, but the course of astronomy …

In the 1980s, Marcy was researching into the magnetic fields of stars: hardly the most gripping of subjects. He admits: 'My career wasn't even exciting to me – how could I expect it to be exciting to my colleagues or anyone else?'

But the creative muse was poised to strike.

'I remember one day – as a post-doc in Pasadena – taking a shower in my apartment,' Marcy recalls. 'I thought to myself that I should try to find a research project that addresses a question I had asked myself as a child. And the question that popped into my mind – with the shower still running – was whether or not there were planets orbiting the stars we see at night.'

That thought has haunted scholars for centuries. In 1584, the renegade monk Giordano Bruno – later to be burned at the stake – asserted, 'there are countless suns and countless earths all revolving around suns in exactly the same way as the planets of our system.'

And Bruno even nailed the reason why it would always be difficult to find these. 'We see only the suns because they are the largest bodies and are luminous, but their planets remain invisible to us because they are smaller and non-luminous.'

The difference was that Marcy had powerful telescopes and spectroscopes to probe the heavens. Instead of looking for the planets directly, his team would scrutinise the parent star, to see if it was wobbling towards and away from us, as the planet circled around it.

'It's a bit like a dog owner with a tiny poodle on a leash,' explains Marcy, where the 'leash' between the massive star and the diminutive planet is gravity. 'Even if you didn't see the poodle, you might see the owner being jerked around by the dog.'

Marcy and his team had to push technology to the limit, to measure how these tiny wobbles affected the star's light. And that wasn't their only problem. 'In the early 1980s,' Marcy recollects, 'the idea of searching for planets was akin to looking for little green men with your satellite dish. It was like pyramid power. People frankly laughed at the idea.'

Undaunted, Marcy set out on his mission: to pioneer the discovery of planets around other stars. It took eight years to perfect his instrument, on a telescope at the Lick Observatory, perched high above Silicon Valley. And he didn't expect success overnight.

To put this in perspective, suppose you were a distant alien, intent on watching how the Sun responds to the gravitational dance of its planets. You'd be in for a long wait. Small planets – like our Earth – have little impact on its vast mass. The only 'poodle' that's tugging on our star's progress through space in any measurable way would be Jupiter. And it takes a whole twelve years to travel once around the Sun.

So Marcy was content to observe his target stars, and save the data for years and years on the computer, until he thought he could pick up this very slow waltz.

Oops, bad call. Because, in the meantime, a rival team came into the frame.

'It was early 1990,' says Didier Queloz, 'when I started my thesis.' He was a graduate student, working with the established astronomer Michel Mayor of the Geneva Observatory. It took them a while to get their equipment working at the Observatoire de Haute Provence, in France.

Like Marcy's team, Mayor and Queloz were monitoring stars, looking for a regular change in speed that might be the work of a planet on the end of a gravitational leash. Mayor didn't expect instant results, either. He booked a six-month sabbatical in Hawaii, leaving his student in charge of the project. Single at the time, Queloz spent his nights with the telescope, just checking how the measurements were coming on. It was a decision that would catapult him to the headlines.

One of the regular targets was a faint star in the constellation Pegasus, a star-pattern that the Greeks saw as a flying horse. The star doesn't have a name, as such, but was listed in the catalogue as 51 Pegasi.

'Every night I was observing that star, I got a different measurement for its speed,' says Queloz. 'It was very, very worrisome.'

Queloz worked out the star seemed to be swinging back and forth in just four days. If this was the work of a planet, it was just crazy. Even the closest planet to the Sun, Mercury, takes almost three months to go around once.

'If I had been a bit older with a lot of experience in the planet field,' admits Queloz, 'I would have said – Oh no, it cannot be a planet.'

But Queloz was filled with the innocence and excitement of youth. He went on to calculate that the planet was not only much closer to its star than minuscule Mercury, but it was also a massive world – fully half as heavy as the Solar System's giant, Jupiter.

This new planet broke all the rules. Queloz remembers 'At the end of every night, I said – OK, maybe I've made a big mistake and I have to sleep. I was very afraid.'

After Mayor returned, they checked and rechecked – and eventually went public at a press conference. 'I'm pretty sure most of the people in the room didn't believe us,' says Queloz. 'But some people thought OK – but ... we'll have to see if somebody else could confirm it.'

The confirmation didn't take long. Marcy's team were in pole position in California. 'We were so shocked by the news that we didn't believe it was a planet at first,' he admits. As soon as they could, they checked out 51 Pegasi for a period of four nights.

His colleague Paul Butler adds, 'When that fourth night of data came in, it did exactly as predicted. We both felt shivers ...'

Astronomers have nicknamed the planet Bellerophon, after the mythical hero who tamed the flying horse Pegasus. It was the first example of a whole new kind of beast that exists in the universe: a 'hot jupiter' is a giant planet that's skimming just above the broiling surface of its sun.

Marcy and Butler went back to the stack of measurements they'd accumulated over the years, and immediately found two more massive planets. 'We'd been working for ten years solid, seven days a week,' says Marcy. 'And finally we looked at our data and there was the wobble. To see that on the computer screen was by far the most momentous moment in my entire life.'

His dream in the shower had come true. And there was a deluge to come. 'It was like a gold rush. Now we knew where to look for planets, and that some might be in fairly close to their star. We looked and – sure enough – we found dozens of them.'

Over the past twenty years, astronomers have tracked down hundreds of hot jupiters, some of which make even Bellerophon look tame.

Lying in the constellation of Vulpecula, the Little Fox, the planet HD 189733b is truly a planet from hell. Though it's a brother to cream-coloured Jupiter and Saturn in our Solar System, HD 189733b is a deep blue, like a vast gaseous sapphire.

The planet's cerulean colour is caused by tiny drops of molten glass, in clouds high in the atmosphere of a world that's blow-torched by its nearby sun. The noonday temperature soars to 1100°C, as hot as the incandescent core of a blast furnace.

This superheated 'air' rushes round to the planet's night side, in winds that roar at 10,000 kilometres per hour. That's a hundred times faster than a hurricane on Earth, and so rapid that the wind is travelling at supersonic speed!

Under its star's intense heat, the atmosphere of HD 189733b is actually boiling away into space: if we could get close, we'd see a cloud of gas steaming away from the planet, and forming a tail like a comet. Planet HD 189733b is losing a million tonnes of its atmosphere *every second!*

Another weird hot jupiter is TrES-2b. This world, orbiting a star in the constellation of Draco – the Dragon – is the blackest planet known. In fact, it's darker than anything else we know in the universe, except for a black hole. Astronomers detected this planet from its silhouette moving in front of its sun. It's another world with superhot air, this time probably filled with atoms like sodium or potassium, which absorb all the light falling on them.

Other early additions to the planetary menagerie included planets that refuse to follow a circular orbit around their suns. 'We found a planet around the star 70 Virginis,' says Marcy, 'that was a real weirdo.' Its furthest point from its sun is twice as far out as its nearest point – which is like the Earth swinging from the vicinity of Venus out to Mars. 'The orbit was so elongated that people even thought to themselves – is this really a planet, or is it some other kind of beast?'

These first discoveries showed that other planet families didn't have to be like our neat and orderly Solar System, with the planets nested in circular orbits and the giants farthest out. Astronomers swung round to the view that maybe *we* are the odd ones out. Planets are born from a swirling disc of gas, dust and ice surrounding a newly born sun; and, in our Solar System, the planets have stayed in much the same place they were born. But perhaps that's not the norm.

Just down the road from Geoff Marcy, at the University of California, Santa Cruz, the new discoveries excited Doug Lin. He had been struggling for years with the theory of how the planets of the Solar System had been born.

'I recalled some of the work done in the eighties on the formation of giant planets such as Jupiter,' says Lin. His theory showed that a disc of material would drag on a planet like Jupiter, so that it would tend to spiral inwards towards its parent star. 'So the discovery of hot jupiters caused me a great deal of joy.'

But it's bad news for any planet that's closer to its sun. A hot jupiter is a killer. Though it's born in the cool outer reaches of a burgeoning system of planets – where gas is rich and plentiful – the giant planet rolls inwards. As relentless as a bulldozer, the hot jupiter sweeps up smaller worlds – like the Earth – and plunges them into a fiery demise in their sun.

And what of the planets in crazily elongated orbits?

'Whenever planets form, they tend to form in a family,' Lin explains. 'And, as you know, in a family, siblings tend to perturb each other. And sometimes this interaction becomes so strong it can break up the family.'

It's the same in a planetary family. As the young planets grow, their gravity strengthens – and they can end up in a wild dance that throws them into strange orbits. 'Sometimes it can be so totally disruptive,' adds Lin, 'that some of the planets may even get kicked out from the neighbourhood of their star.'

Sure enough, astronomers have recently been finding 'orphan planets', ejected from the home system and now wandering alone in deep space. Some eighty light years from us, in the constellation Capricornus – the Sea-Goat – swims the planet PSO J318. This poor world is not only saddled with an unwieldy catalogue name, but is a celestial orphan. It has no sun. With no warming rays to heat its forever gloomy surface, PSO J318 is doomed to a lifetime freezing in the ultimate icebox.

There may, in fact, be more frozen orphan planets roaming our Galaxy than there are stars …

Geoff Marcy and the teams that followed him set out in the hope of finding planets that bore a family resemblance to our own Solar System, containing – perhaps – a small planet like Earth which would be a heaven for future space-farers. Instead, they found planets from hell.

What went wrong?

'Maybe the first gold nuggets we are picking up today are the oddballs,' Doug Lin presciently said a few years after the first planets had been discovered. 'With current technology, we are finding systems that are the easiest to detect – with planets that are relatively massive. As we perfect our observational techniques, we'll find planets with lower mass. These systems will be very stable – they'll resemble our Solar System once more.'

In human terms, it's like going to a party full of strangers. At first, you'll notice the larger-than-life characters, the loudest, the weirdest. After a while, though, you realise that most people there are much more normal.

Lin turned out to be absolutely right. But astronomers needed a new technique, to winkle out the smaller and more normal planets. Enter Dave Charbonneau. Like Didier Queloz a few years earlier, he was a young graduate student in search of a project. Between them, Charbonneau and his advisor Bob Noyes came up with a plan.

'At the time there had been dozens of planets detected by the wobble technique,' says Charbonneau. 'And we wanted to come up with an idea that would allow us to learn something else about these planets.'

Charbonneau and Noyes had a 'eureka' moment. 'If the planet passes in front of the star, you'll see a little dimming of the light from the star for about three to four hours,' Charbonneau explains. Astronomers call these events 'transits': you may remember that the planet Venus made a transit of the Sun a decade ago, when it appeared as a tiny black speck against the Sun's brilliant face.

In the case of a distant star, you can't actually see the planet as a silhouette; but you can measure how much it dims the star's light. 'From the dimming,' Charbonneau continues, 'you can measure the size of the planet, and its mass – these two things together would tell you the density: what the planet's made of.'

From Michel Mayor in Geneva, Charbonneau learned that a star in the constellation Pegasus – not far from the first discovered planet, Bellerophon – had a planet circling in a very tight orbit. This star was fairly bright, so Charbonneau didn't need a huge telescope. In fact, he observed its brightness with a telescope a mere 10cm in diameter – only three times the size of Galileo's largest telescope!

'When I saw the first indication of a dimming,' enthuses Charbonneau, 'it was really, really exciting. My first reaction was "it's too good to be true".' So he checked another night's data. 'When in those data I saw the transit emerge once again, I had to sit back in my chair.'

Time to do the sums, to work out the planet's size. 'Although its mass was similar to Jupiter,' says

Charbonneau, 'it turned out to be significantly larger. It had been puffed up due to the proximity to the star and all the heat that was hitting it.'

Here was a great new way to discover planets, too. If you could stare at a huge number of stars all at once, you could check out which ones were subtly 'winking' at you. And there's an added bonus: it's easier to measure a star's brightness than to check out the tiny wobble caused by a planet on a gravitational leash, so you can hope to find smaller worlds.

Without being too unromantic, we have to take a moment here to curse the 'twinkling' of the stars, which makes it difficult to detect the faint dimming when an orbiting planet transits the star. There is a way out, though. Twinkling is caused by starlight passing down through the Earth's unsteady atmosphere: observe from space, and the stars no longer twinkle.

That's why, in March 2009, a rocket blasted off from Cape Canaveral in Florida, carrying a spacecraft called *Kepler*. Named after the astronomer who worked out how our local planets orbit the Sun, it was designed to seek thousands of planets circling other stars.

Kepler carried a telescope that's pretty small in the grand scheme of things, less than half the size of the mirror in its big brother, the Hubble Space Telescope. But *Kepler* wasn't designed to take pretty pictures. Its task was to stare continuously at a patch of sky in the constellation Cygnus, carefully measuring the brightness of over 100,000 stars.

'*Kepler* has undoubtedly revolutionised the field of exoplanets,' enthuses Elisa Quintana, who has been with the project since the beginning. 'I thought it was fascinating that there could be other worlds out there.

I also got more interested in space around that time because my physics advisor was Dr Sally Ride, the first American woman astronaut that flew in space.'

In four years – until part of its control system failed – *Kepler* pinned down 995 planets circling other stars: astronomers like Quintana are still checking out many other candidates. *Kepler*'s eagle eye has picked out planets far smaller than Jupiter: in fact, some of the tiniest worlds known to exist outside the Solar System.

'*Kepler* has been so important in showing us that exo-solar systems come in many flavours,' says Quintana, 'and the diversity is astounding. There is no "one size fits all" model.'

In particular, *Kepler* has discovered planets unlike any in the Solar System. In our planetary family, we have rocky planets no larger than Earth; and gas giants with diameters bigger than Neptune, which is four times wider than our planet. There are no planets intermediate in size. But it's different out there …

'*Kepler* has found that super-Earths and mini-Neptunes are abundant,' Quintana says. 'These are planets of which there are no analogues in our Solar System! In fact, these are the most abundant planet sizes that *Kepler* has found so far.'

In the case of Kepler-11, we have six 'super-Earths' orbiting the star in a region hardly larger than Mercury's path around the Sun. 'It really shows how compact planetary systems can be,' says Quintana.

So – let's visit some of *Kepler*'s new worlds …

For keen surfers, there's the planet Kepler-62e. It's some 60 per cent bigger than the Earth, and totally inundated with oceans, thousands of kilometres deep. Though water is an essential ingredient of life, this waterworld is not a place where intelligent aliens could evolve, according to Geoff Marcy: 'the oceans would be so deep that they would cover the continents, cover the mountains and there would be no dry land on which to design grand pianos, great art, literature and of course computers and rocket ships.'

If you're a pyromaniac, try out Kepler-78b. Though it's a rocky planet and the same size as the Earth, this world is a natural inferno. The planet's tight orbit gives it a 'year' lasting only eight hours. The side facing its sun is a roiling sea of molten lava, broiled to a temperature of almost 3,000°C – like being continuously blasted by the flame from an oxy-acetylene welding torch.

Romantics should head for Kepler-16b. This planet is orbiting around a pair of stars, so you'd witness two sunsets every day. Elisa Quintana wrote her thesis on how a planet could form in a binary star system, so 'when *Kepler* started detecting circumbinary planets, I was thrilled!'

But she also has a soft spot for Kepler-37b, the smallest exoplanet yet found – even tinier than Mercury in our Solar System.

With *Kepler*'s rich bounty of planets, astronomers can now see that the Solar System is not such an oddity after all. There are plenty of other suns with a sedate family of planets, following orderly paths year after year.

Kepler's census of planets, large and small, allows astronomers to work out the chances of finding a 'second Earth'. The result is astounding: more than 20 per cent of stars like the Sun should have an Earth-twin in the habitable zone, where the temperature is just right for water to be liquid. In our Milky Way Galaxy, that means eight *billion* planets like our Earth, basking in the light of a Sun lookalike. '*Kepler* has for the first time in human history told us that our planet is not unique,' says Marcy. 'It's not even rare.'

And that's not all. Fainter 'red dwarf' stars are far more ubiquitous than the Sun, and they could have planets just as homely as the Earth. The main difference is that the red dwarfs burn less fiercely, so a planet must huddle closer in to keep warm.

It would be easy to dismiss these miserable specimens of stardom; but Marcy points out: 'It could be that most of the abodes of life in the Galaxy are actually planets orbiting these tiny little red dwarf runts.'

The statistics suggest there are eleven billion habitable planets orbiting red dwarfs in our Milky Way – in addition to the eight billion that have a Sun-like parent star. Any of these could be home to life like ours; and even the abodes of intelligent aliens that may be trying to contact us (as we'll explore in the next chapter).

In 2014, Elisa Quintana led a team that tracked down the first true twin to the Earth: its name is Kepler-186f.

'It was definitely an exciting journey,' she recalls. 'We find many planet candidates that look to be small and in the habitable zone, but each system requires careful study.' The more they studied Kepler-186f, the more Earth-like it looked. 'It was fun – and a bit scary as well – as we were about to provide a proof-of-concept that other Earth-size planets exist in the habitable zones of other stars.'

Kepler-186f is a member of a family of planets orbiting a red dwarf star that lies 500 light years away in the constellation Cygnus (the Swan). The Earth-twin is orbiting at about the same distance as Mercury spins around our Sun, where it soaks up enough heat from the feeble star to keep its climate mild.

Relaxing after the announcement, Quintana admits that planet-hunting isn't all hard work. 'There are days that are just fun,' she grins. 'While working on the Kepler-186 system, we were lucky to have some wonderful artists creating some illustrations for the system. I just thought to myself, I can't believe that my job is to sit at a desk and imagine what other worlds might look like!'

Quintana's Holy Grail is still to find a planet like the Earth orbiting not a red dwarf, but a star more like the Sun. 'As we have lots of *Kepler* data that's still being analysed,' she says, 'I'm hoping we will find one!'

And she looks forward to new space telescopes that will extend the search. The Transiting Exoplanet Survey Satellite, due to launch in 2017, can seek out planets over the entire sky. The following year's James Webb Space Telescope – the mighty successor to Hubble – will check out the atmospheres of these new worlds, looking for signs of primitive life: 'I hope that the detection of life outside of our Solar System happens within my lifetime,' Quintana concludes.

Even after two decades, the search for 'worlds beyond' has lost none of its excitement. Pioneer Geoff Marcy admits that his full-on years of winkling out planets hasn't been just a job, but a vocation. 'Part of it is my ego, actually. And it's a little embarrassing to say, but I would love to make a contribution to science. I love science. I love human knowledge of any kind.'

The visceral excitement of the planet-chase was perhaps best expressed not by a scientist, but by a poet. When John Keats opened a volume of poems from the ancient Greek Homer – as earthily translated by an Elizabethan poet – he was inspired to pen a sonnet, *On First Looking into Chapman's Homer*.

With the explorers of America in mind, and William Herschel's discovery of Uranus only a generation before, Keats concluded his poem with lines that would resonate with Marcy, Quintana and all of today's other dedicated planet-hunters:

> Then felt I like some watcher of the skies
> When a new planet swims into his ken;
> Or like stout Cortez when with eagle eyes
> He star'd at the Pacific – and all his men
> Look'd at each other with a wild surmise –
> Silent, upon a peak in Darien.

Chapter 18

ARE WE ALONE?

Nearly five centuries before the birth of Christ, the Greek philosopher Anaxagoras made an astounding assertion regarding what was happening on other worlds:

'Men have been formed and other animals which have life; the men too have inhabited cities and cultivated fields as with us; they have a Sun and a Moon. And their Earth produces for them many things of various kinds, the best of which they gather together into their dwellings and live upon.'

In other words, Anaxagoras believed in alien life. And – two and a half millennia on – so does astronomer Jill Tarter, holder of the Bernard M. Oliver Chair at the SETI Institute in California. SETI is an acronym for one of the greatest enterprises that the human race has ever undertaken: the Search for Extraterrestrial Intelligence.

On 12 October 1992 – Columbus Day in America – Jill had a momentous task to perform. We were privileged to be with her at Arecibo in Puerto Rico, dazzled by the sight of the world's biggest radio telescope, nestled in the greenery of the island's jungles. The giant dish – twenty-six football fields in size – had been tuned to listen in for whispers from intelligent life in the universe.

We were about to celebrate the official start of humankind's ultimate quest. Jill flung the switch on the cosmic leviathan, and proclaimed the immortal words: 'Let the Search commence'.

The first data began to pour in, covering millions of radio frequencies being emitted from hundreds of stars.

Jill considered the nature of what she and her team were embarking upon. '"Is anybody out there?" It's the oldest unanswered question our species has posed to itself.'

Tarter is passionate about SETI. 'What's so terrific is that we suddenly have the technology that allows scientists and engineers to answer the question as to whether there's life in the universe. For me, I can't imagine doing anything more important.'

She follows in the footsteps of the 'father of SETI', Frank Drake. He recalls when the instinct to search for extraterrestrial life first hit him. 'During my doctoral thesis at Harvard in the fifties, I became intrigued with the possibility that we might really be able to find ET. But – back then – the ideas of life in the universe were not considered very reputable subjects in science.'

Young Drake was not deterred by the orthodoxy. Instead of using a conventional optical telescope, he surveyed the heavens with a giant radio dish, designed to look for exploding stars and violent galaxies.

Radio waves travel through space virtually unimpeded. Could extraterrestrials, he wondered, use a radio telescope 'in reverse' to *broadcast* signals across the cosmos? After all, we on planet Earth use radio to make broadcasts 24/7.

His bold vision seemed to have come true when he was observing the beautiful Pleiades star cluster. Drake describes what happened: 'I'd done this many nights before. But on this occasion, there suddenly appeared a very strong narrow-band signal, which could only be the product of intelligent activity.'

To check if the signal really was coming from the Pleiades, Drake slewed his dish to a different part of the sky. 'Well: it turned out that when I moved the telescope, the signal was still there. So it was truly from Earth – which was a disappointment. But the seed was planted.'

And the seed bore fruit. Many other radio astronomers eagerly joined the search. Among them was Bob Dixon from Ohio State University. Using a radio telescope called 'Big Ear' – constructed by students – he spent twenty years listening in for the call from ET, before the message arrived.

'It came in 1977,' Dixon tells us. 'It's come to be known as the "WOW!" signal, because it was exactly the kind of signal that we'd been looking for: it was narrow-band, coming from a long way away. It had all the hallmarks of intelligence. The person who was looking at it at the time got so excited that he wrote WOW! on the computer print-out – and hence that name has stuck.'

Dixon continues: 'Unfortunately, it was only there once. We saw it for about a minute; in fact, it turned off while we were watching it. We went back there hundreds of times later to look for it, and it was never there.'

So what was it?

'Well – it's not something on the Earth,' Dixon ventures. 'It's not something in Earth orbit, because it would have looked different on our chart records at the time. It *could* have been some other civilisation. Or it could have been some super-secret military satellite cruising around the Solar System somewhere. But they will never admit to the existence of that – so we'll never know.'

And Bob Dixon will never get to hear any more alien messages from the cosmos. In its infinite wisdom, the local authority bulldozed 'Big Ear' – to make way for a golf course.

SETI researchers are used to setbacks like this. But it hasn't deterred them because there's been a growing conviction that life in the universe must be commonplace. We already know of nearly 2,000 planets orbiting other stars. And the chemicals that make life – in particular, carbon – are widespread in space. Put together these facts, and it seems very unlikely that we are alone.

Astrobiologists have recently been discovering just what extremes life on Earth can tolerate. There are bacteria that can live in nuclear reactors, at the bottom of deep bore-holes, in the cold of Antarctica, and in the superheated pools of Earth's hot springs. These 'extremophiles' can even survive on pallets exposed to the vacuum of deep space.

The lesson's clear: life – once established – clings on … for dear life.

So, could there even be extreme life elsewhere in our Solar System? Mars has always seemed the best bet, as we've described in Chapter 16. But there are other candidates.

Centre-stage is Europa, one of the larger moons orbiting Jupiter. It has a smooth icy surface, but, because Europa is being gravitationally pummelled by Jupiter, it must be warm inside – which leads researchers to believe that the moon has a deep ocean under its icy crust. In Arthur C. Clarke's 1982 novel *2010: Odyssey Two*, a space crew from Earth encounters watery lifeforms from Europa's deeps.

And then there's Titan, the largest of Saturn's moons, cloaked in orange clouds that float in an atmosphere of nitrogen twice as dense as the Earth's. In January 2005, the European *Huygens* probe landed on its surface. It found a world quite unlike anything else in the Solar System. Though intensely cold, Titan has lakes of liquid methane or ethane, plus hotspots of volcanic activity. Taken together, these are the basic conditions to create life.

But even if there's life elsewhere in the Solar System, it's going to be pretty primitive. Beyond our local neck of the galactic woods, could there be not only life, but *intelligent* life?

The discovery of planets orbiting other stars was a real spur to the SETI astronomers. But there were trials and tribulations along the way. The NASA-funded search at Arecibo in 1992 was dead just a year later, killed off by a politician who thought it frivolous. The person responsible for cutting off the funding – just $100 million spread over ten years – was a Nevada senator called Richard Bryan, who derided the project as 'a great Martian chase'. Astonishingly, he also managed to persuade his colleagues in the US Congress to support him.

There was an enraged reaction from the press, but nothing could change the situation. 'While grocery-store tabloids scream about flying saucers, NASA is looking for the real thing,' lamented the *San Francisco*

Examiner. The *Boston Globe* rued: 'It proves one thing, and one thing only. That there is no intelligent life in Washington.'

That could have been the end of the story. But Tarter, Drake and the rest of the team were made of sterner stuff. They transformed themselves into business-people, created a private company, and set about an international fundraising campaign. After fifteen months, the team had several million dollars in the bank – and they were also bound for Australia to conduct the first southern hemisphere SETI search on the Parkes radio telescope in New South Wales. Project Phoenix – with Jill Tarter at the helm – had been born out of the ashes.

And now they have the beginnings of their own new radio ear on the cosmos. At Hat Creek in Northern California, a clutch of six-metre dishes is springing up. Funded by Paul Allen – co-founder of Microsoft – the completed Allen Telescope Array will grow to 350 dishes. Jill Tarter is upbeat about the array. 'We now have forty-two dishes', she tells us.

The new SETI search is called SonATA (SETI on the Allen Telescope Array). 'We search three exoplanet systems simultaneously – mostly from the Kepler field,' says Jill. These are planets in Cygnus, picked out by NASA's *Kepler* spacecraft, as we've described in the previous chapter. She adds: 'We're also searching all the confirmed exoplanets from ground-based observations as well.'

But over half a century after Frank Drake's wake-up call – when he thought he'd detected a signal from the Pleiades – the team still hasn't received a phone call from ET.

Maybe aliens are simply strong but silent. After all, humankind has done little to advertise its presence to the universe at large. Drake, however, recalls some of our previous, somewhat misguided, attempts to communicate our existence: 'It turns out that many brilliant people in the past have thought about contacting ET, and have even come up with schemes which they thought were excellent ways to proceed.

'One of the earliest was the great mathematician Karl Friedrich Gauss,' Drake continues. 'In the 1820s, he proposed that we should plant – in Siberia – a right-angled triangle filled with wheat. And on each side of the triangle, he proposed to plant a square of pine trees demonstrating that the sum of the squares on two sides equalled the square on the hypotenuse.' A giant expression of the famous Pythagoras's Theorem that we learn at school, it would be visible to other creatures in the Solar System, and would prove the existence of intelligent life on Earth.

Twenty or so years later, the Viennese astronomer Joseph von Littrow came up with an even more ambitious scheme. Relates Drake: 'It was to dig trenches in the Sahara Desert in the form of geometrical figures – squares, triangles, circles – and then fill them with kerosene. In the dark of night, these ditches filled with flammable fluid would be set on fire, creating blazing geometrical figures recognisable all the way to Mars, Venus and beyond.'

Drake followed in his predecessors' footsteps, and made his own attempt to say 'Hi!' to the universe. But it was the brainchild of his PA – not his own! Drake was director of the Arecibo Observatory in 1974, when the giant radio dish had just been resurfaced. He wanted to host a ceremony to celebrate the event – and his secretary came up with the brilliant suggestion: 'Why don't you use the dish in reverse and beam a signal to the cosmos?'

Frank Drake recalls the occasion: 'In the middle of the celebration, we transmitted to the stars in the direction of the great globular cluster M13. This is a group of 300,000 stars 25,000 light years away from us. The message was made in a code that was easy to break.'

Drake adds: 'We showed the basic chemistry of life on Earth, the DNA molecule, the arrangement of our Solar System – and we gave our population, Earth's size, and the size of the telescope that sent it.'

But don't hold your breath. The Arecibo message will take 25,000 years to arrive at its destination. And the reply – if any – won't come back for a further 25,000 years. Communication with ET is hardly snappy.

Drake, in collaboration with Carl Sagan – he of TV's *Cosmos* fame – went further. They designed plaques that were attached to the two *Pioneer* spacecraft destined for Jupiter and Saturn, and are now heading out of the Solar System. If any extraterrestrials locate the *Pioneers*, the images on the plaques will teach them about our knowledge of science, our perception of the universe – and even our anatomy.

'The two human figures on the *Pioneer* plaque are both nude,' points out Drake. 'And we thought – the extraterrestrials are interested in just what our anatomy is, which is why the humans are that way.'

But the design of the *Pioneer* plaques did not go down well with everyone. 'This caused a great deal of consternation,' Drake admits. 'In many American newspapers, the drawings were altered to remove any evidence of sex organs from them. Soon after the plaque was launched, I was invited to appear with the plaque on Canadian television. There was anxiety, upset, worry ... because, as it turned out, this was the first time nude human beings had ever been shown on Canadian TV. In fact, nobody complained.'

The citizens of America proved a trifle more narrow-minded. 'In the *Los Angeles Times*, there were a number of letters to the editor, protesting that we were using taxpayers' funds to send "smut into space".'

Later, the twin *Voyager* craft – also aimed at the outer planets, but eventually destined to escape towards the stars – carried a more modest message. Unbelievably, both probes sported an old-fashioned LP record on their sides, complete with a stylus! The discs were encoded with sounds, pictures and messages from Earth.

Robot messengers like the *Pioneers* and the *Voyagers* may be the way to carry our thoughts, our senses and our knowledge into the community of extraterrestrial life – and a first step to going there ourselves. But Drake acknowledges that our communications to extraterrestrial life are more of a reminder of the nature of human life to our own inhabitants. 'Why did we do it? Well – we did it really as a message to Earth.'

When it comes to finding the aliens' putative message to us, the SETI Institute's Allen Telescope Array is not the only show in town. Up the freeway in Berkeley, Dan Werthimer and his team run Project Serendip VI, a SETI search which has a different approach. 'We call it piggy-back SETI,' Werthimer explains. 'We use the Arecibo telescope simultaneously with other astronomers, and we're running twenty-four hours a day all year long at the same time as other astronomers are using the telescope.'

'The other difference is that there are different strategies for looking at radio signals,' Werthimer explains. Project Phoenix was a 'targeted search', which selected a thousand targets – nearby stars that are like our Sun. The advantage was that the SETI researchers could concentrate on them and spend a long time on each star.

'Our strategy is called a sky survey,' Werthimer continues. 'We're looking at billions of stars – as far as Arecibo can see. But we can't spend as long at each place in the sky.'

The Serendip equipment – now in its sixth version – is formidably powerful. It can listen to five billion radio signals simultaneously. Today's SETI relies not only on enormous telescopes, but also on enormous computing power. 'It's computers that do the listening,' explains Werthimer.

Even so, Dan isn't satisfied. He wants even more computer power. 'We're asking people all around the world to help us analyse the data from the Arecibo telescope. Eight million people are participating in this project, and together they will make a megacomputer hundreds of times larger than the world's current biggest supercomputer.

'Everybody gets a little piece of data from the telescope, everyone gets assigned a little bit of the sky. Then they download a free screensaver programme into their PC or Mac, and it analyses the data. After a few days it sends its results back to Berkeley, and the participants automatically get some more data to analyse.'

The SETI@home project, as it's called, could catapult a fourteen-year-old computer geek into the spotlight if he or she happens to turn up an alien message. 'If your screensaver finds the extraterrestrial civilisation, you'll become quite famous. But don't hold your breath.'

Has Werthimer found any tantalising signals? 'We've had a few hundred things that have been interesting enough that we've gone back to the telescope to check them out again – but, so far, they've all turned out to be terrestrial radio signals (from humans and their transmitters), not extraterrestrial transmitters.'

Dan Werthimer isn't putting all his eggs in one basket when it comes to SETI. 'Besides looking for radio signals from other civilisations, we're also looking for laser signals. The idea is that perhaps – instead of radio waves – they may be sending us pulses of light. So, using an optical telescope, we're looking for very short, bright pulses, perhaps a billionth of a second long, that may be brighter than their star.'

When we spoke to Werthimer he went on to confess: 'One of the reasons that we're looking for laser signals is because the guy who invented the laser – Charlie Townes – is in the office right next door to me. And he's been pestering me for years, saying: "Hey! Instead of looking for radio signals, why don't you try looking for laser signals?" So finally, after enough pestering, here we are at the telescope looking for these signals!'

Werthimer was full of admiration for the veteran physicist. 'Charlie comes to work every week, at ninety-nine years. He's a role model for all of us.' Very sadly Townes died in January 2015, just six months short of his one-hundredth birthday.

Dan Werthimer concedes that communication by laser does have its advantages. 'Lasers have very directed, narrow beams – they might be the perfect things for interstellar communication. Another advantage is that you can put a lot of information on a laser beam – perhaps billions of bits per second. You could send your whole Library of Congress in a minute or two.'

And lasers are also incredibly powerful. 'If you take a big laser like the kind that's at the Livermore Laboratories and put it on a telescope like the kind in Hawaii, you can send messages across the Galaxy.'

Are lasers the last word in communications? 'If you'd asked me two hundred years ago how we might communicate with other civilisations, I might have said that smoke signals were the best way. If you ask me now, I'll say radio signals or perhaps laser signals are the best technique. That means that if you ask me two hundred years from now, there might be something even better – perhaps faster-than-light particles called tachyons, or some sort of new particle we don't even know about.'

Whatever the means of communication, Werthimer believes that there is a great deal of information exchange going on in the Galaxy. 'It's possible that other civilisations have been in touch with each other for perhaps billions of years, all talking to each other with laser beams or radio waves. We're an emerging civilisation, just getting in on the game. But we might become part of that galactic internet.'

One day, SETI researchers are confident that they will get the call. Jill Tarter ponders on how we will react. 'What if we get a signal? What will happen? Will you still go to work tomorrow, or will the world stop?'

Adds Dan Werthimer: 'If we do find an interesting signal, our first task is to see if another group can find it and independently verify it – because it might be a bug in our software, or something wrong with our equipment, or a graduate student trying to play a prank on us.

'If it's real, then we'll make an announcement – and that announcement will go out all over the world. It'll be to all people, to all countries, we'll make it available on the Web – all the information will be shared.'

Werthimer is extremely upbeat as to the content of the first message that we receive across the interstellar void. 'If they're sending a message intentionally, we have a lot to learn from them. That means they will make it anti-cryptographic – they'll make it with language lessons that are easy to decode. They'll probably send their music, their poetry, their literature, their medicine, their science. I'd love to learn about their music!'

Frank Drake has – naturally – pondered the question of getting the call many times over the past sixty years. 'There'll be an early period of excitement – you know, just what is this, what are we seeing? The biggest changes, however, will come over decades as we learn about these civilisations – because if there's one, there's more. In most cases, they will have been technologically capable for a much longer time than us, and we will learn things about ourselves, in very important things. Such as what our potential is, what we can become, what we can evolve to.

'Overall, it should create a change in our way of life on Earth which is greater than any that has ever occurred – and one much to our benefit, because it will be mostly good things that we learn, helpful things.'

'I think it's likely that advanced civilisations are going to be peaceful,' adds Werthimer. 'The ones that are not so peaceful are going to blow themselves up, and they're not going to be around any more. I don't think other civilisations are going to come and eat us. And I don't think they're killing each other the way we are.'

The other big question is: should we reply to the message – and *who* should reply on behalf of Earth? The SETI researchers currently have a protocol drawn up, which would involve a major decision on behalf of the whole of humanity.

Jill Tarter flags up the protocol website. It points out firmly: 'No response to a signal or other evidence of extraterrestrial activity should be sent until appropriate international consultations have taken place.' It adds: 'Signatories to this declaration will not respond without first seeking guidance and consent of a broadly representative international body, such as the United Nations.'

But – in the end – the question is a no-brainer. If there *are* extraterrestrials out there, they already know about us. Radiation from our radio and TV stations – not to mention powerful military radars – does not confine itself to planet Earth. We have been inadvertently transmitting radio signals from our world in their untold millions, every day, for almost a century. 'We have made ourselves brilliantly conspicuous to the universe,' says Frank Drake.

Drake isn't at all worried about the threat from aggressive aliens being aware of our existence. Speaking for the SETI community, he declares: 'We all stay dedicated to SETI – working hard for it – because the detection of extraterrestrial life is the most exciting thing we know of to do. And we want it to happen. Not just for ourselves, but for the whole of humanity.'

'Maybe we're alone, maybe we're not – either way, the answer is really significant,' muses Jill Tarter.

Dan Werthimer looks forward: 'I'm optimistic in the long run – probably in our lifetimes – that we will be in contact with other civilisations.

'It would be bizarre if we were alone. It would be a cramped mind that didn't wonder what other life is out there.'

INDEX